Emerging Trends in Fisheries - Sustainable Practices and New Perspectives

Edited by Muhammed Atamanalp, Momin Momin, Gonca Alak, Arzu Uçar and Veysel Parlak

Published in London, United Kingdom

Emerging Trends in Fisheries - Sustainable Practices and New Perspectives
http://dx.doi.org/10.5772/intechopen.1006014
Edited by Muhammed Atamanalp, Momin Momin, Gonca Alak, Arzu Uçar and Veysel Parlak

Contributors
Arzu Uçar, Ashraf M.A. Goda, Catherine Maindi, Cengiz Kılıc, Chrysantus Tanga, Divine Dogbeda Azumah, Ehab El-Haron, Einar Ringø, Erdi Aksakal, Esau Matthews Mbokane, Gonca Alak, Hlulani Archiebold Hlungwani, Isaac Osuga, Mohamed Ashour, Mostafa A.M. Soliman, Muhammed Atamanalp, Nuray Karaca, Seth Koranteng Agyakwah, Shadrack Kwadwo Amponsah, Veysel Parlak, Vincent Mwashi

First published in London, United Kingdom, 2025 by IntechOpen
IntechOpen is the global imprint of INTECHOPEN LIMITED, registered in England and Wales, registration number: 11086078, 167-169 Great Portland Street, London, W1W 5PF, United Kingdom

For EU product safety concerns: IN TECH d.o.o., Prolaz Marije Krucifikse Kozulić 3, 51000 Rijeka, Croatia, info@intechopen.com or visit our website at intechopen.com.

British Library Cataloguing-in-Publication Data
A catalogue record for this book is available from the British Library

Emerging Trends in Fisheries - Sustainable Practices and New Perspectives
Edited by Muhammed Atamanalp, Momin Momin, Gonca Alak, Arzu Uçar and Veysel Parlak
p. cm.

This title is part of the Agricultural Sciences Book Series, Volume 25
Topic: Animal Farming
Series Editor: W. James Grichar
Topic Editor: Vasileios Papatsiros

Print ISBN 978-1-83634-203-8
Online ISBN 978-1-83634-202-1
eBook (PDF) ISBN 978-1-83634-204-5
ISSN 3029-052X

If disposing of this product, please recycle the paper responsibly.

IntechOpen Book Series
Agricultural Sciences
Volume 25

Aims and Scope of the Series

The importance of agriculture cannot be overstated. It helps sustain life, as it gives us the food we need to survive and provides opportunities for economic well-being. Agriculture helps people prosper around the world and combines the creativity, imagination, and skill involved in planting crops and raising animals with modern production methods and new technologies. This series includes two main topics: Agronomy and Horticulture, and Animal Farming. This series will help readers better understand the intricacies of production agriculture and provide the new knowledge that is required to be successful. The success of a farmer in modern agriculture requires knowledge of events happening locally as well as globally that impact input decisions and ultimately determine net profit.

Meet the Series Editor

W. James Grichar has been employed with Texas A&M AgriLife Research for over 45 years with an emphasis on research in agronomy, plant pathology, and weed science. He obtained his BS from Texas A&M in 1972 and his Masters of Plant Protection in 1975. He has published 195 journal articles, over 330 research reports and briefs, 11 book chapters, and over 300 abstracts of profession meetings. He also directs research in many crops including corn, grain sorghum, peanuts, and sesame. He has held various positions in different professional societies including the American Peanut Research and Education Society, Southern Weed Science Society, and Texas Plant Protection Conference in addition to being Associate Editor for Peanut Science and Weed Technology. Significant accomplishments have included spearheading efforts to determine the optimum planting time for soybean production along the upper Texas Gulf Coast. These efforts have shown growers that soybean yields can be improved by 10 to 20% by following a late March to early April plant date. He also has been instrumental in developing a herbicide program for peanut production in the south Texas growing region. Through the development and use of herbicides that are effective against major weed problems in the south Texas region, peanut yields have increased by 25 to 30%.

Meet the Volume Editors

Prof. Dr. Muhammed Atamanalp has been a faculty member at Atatürk University's Faculty of Fisheries since 2009 and has served as its founding Dean. He has nearly 130 SCI papers, most of which are on aquatic toxicology. He is a specialist in water pollution, fish physiology, and the toxic effects of newly synthesised compounds on aquatic organisms. He holds seven registered patents. He has supervised 10 Ph.D. students and 14 master's students and completed 30 fully funded scientific projects. He also serves as a jury member for academic recruitment at over 25 universities.

Dr. Gonca Alak is a professor at Atatürk University's Fisheries Faculty, working on seafood processing and aquatic toxicology with dozens of research projects. Her research area of specialization is seafood processing technology and aquatic toxicology. She works with analytical chemical methodologies to detect bioactive compounds, food contaminants, and packaging migrants. In recent years, it has also been exploring natural extracts from plants and food by-products for integration into active packaging systems.

Arzu Uçar is a Professor at Atatürk University's Fisheries Faculty at the Department of Aquaculture in Erzurum,Türkiye. Dr. Uçar's research spans the fields of aquatic toxicology and the environment, integrating molecular approaches with organismal and population-level measurements. Dr. Uçar's primary focus is the effects of stressors such as emerging pollutants, plastics, and a changing climate on aquatic organisms, but his research and teaching also encompass the connections between ecological and human health.

Veysel Parlak is an Associate Professor in the Department of Aquaculture, Faculty of Fisheries at Atatürk University in Erzurum, Türkiye. Dr. Parlak's research encompasses the areas of aquatic toxicology and the environment, integrating molecular approaches with organismal and population-level measurements. Dr. Parlak's primary focus is water pollution, fish physiology, and aquatic toxicology. His research and teaching also encompass the links between ecological and human health.

Momin Momin is an Assistant Professor in the Department of Aquaculture at the Faculty of Fisheries, Atatürk University, in Erzurum, Türkiye. He holds a Ph.D. from Istanbul University. His research focuses on fish physiology, gamete quality, and aquatic toxicology. Dr. Momin has published numerous articles in prestigious, high-impact journals and has completed several fully funded scientific projects. He actively participates in the international academic community, having attended and presented at numerous conferences, congresses, seminars, symposiums, and workshops worldwide.

Contents

Preface

The global fisheries sector is transforming as we seek innovative and sustainable solutions to the challenges facing aquaculture and aquatic ecosystems. *Emerging Trends in Fisheries – Sustainable Practices and New Perspectives* extensively analyzes the interactive dynamics of sustainable practices, aquaculture, and rural economies.

The book covers a wide range of emerging trends, including the contribution that family-operated aquaculture enterprises can make to address rural-to-urban migration, exploring innovative feed technologies like insect meals as nutrition sources, using seaweed in integrated mariculture, and the critical importance of plankton in marine larval nutrition for fish and shellfish. Additionally, the book explores the untapped potential of inland fisheries, particularly in regions like South Africa, where they hold promise for strengthening rural economies. Each chapter highlights a key component of the new landscape of the fisheries sector, providing valuable information on how to transition toward sustainable development and healthier ecosystems.

I hope that this book will inspire further research, spark new ideas, and contribute to the development of sustainable practices that will shape the future of the fisheries sector.

Dr. Muhammed Atamanalp
Professor,
Atatürk University Faculty of Fisheries,
Erzurum, Türkiye

Dr. Gonca Alak, Dr. Arzu Uçar, Dr. Veysel Parlak and Dr. Momin Momin
Atatürk University Faculty of Fisheries,
Erzurum, Türkiye

Chapter 1

A Strategic Solution to Curb Rural-to-Urban Migration: Aquaculture-Based Family Businesses

Nuray Karaca, Erdi Aksakal, Cengiz Kılıc, Arzu Uçar, Veysel Parlak, Muhammed Atamanalp and Gonca Alak

Abstract

Global environmental problems have led societies to seek more sustainable production and consumption methods in every social field. Although the food sector has contributed significantly to global environmental issues, there has been little interest in terms of research. Family businesses have an important share and contribution in the world economy. With this definition, this study was conducted on family-run enterprises in aquaculture in Erzurum and its districts. Our aim is to demonstrate that the success of family business enterprises in ensuring continuity can be used as a method to prevent migration in rural agricultural areas. Study has been underlined that aquaculture farmers continue their family businesses because their work satisfies them economically, but it is not possible to create continuity in this type of business without state support in unfavorable economic conditions.

Keywords: rural area, village sociology, migration, fisheries, family business

1. Introduction

Over the past century, despite the marked growth in food production, nearly 900 million people do not have access to enough proteins, carbohydrates, and lipids that provide the energy necessary for their metabolism. In this scenario, fishing is rapidly decreasing due to declining existing fish stocks, while aquaculture has been increasing over the past 20–30 years, creating great potential to ease the pressure on natural stocks. Fish farming is the fastest growing sector in agriculture with an annual growth rate of close to 6%, and according to FAO, fish and fish products are an important source of protein globally, especially in low-income countries. In addition, aquaculture production needs to increase by 20% to meet the expected per capita demand for fish in 2030 [1]. Besides, the phenomenon of rural-urban migration in Turkey and the socioeconomic infrastructure that constitutes migration are closely related to the developments in the agricultural sector. As in the whole world, the socioeconomic, sociocultural structure has been greatly affected by population movements, wars, and the problems caused by them in our country [2]. Approaches to rural development and change in Turkey have mostly continued to be influenced by Western education. In parallel with industrialization, the studies defined in the field of rural sociology emerged primarily in the West. It is thought that this study will contribute to the permanent policies that can be implemented on the subject. Internal migration, which is generally attributed to the repulsion of the countryside and the attraction of the city, has various reasons that differ from country to country [3]. The rural-urban dichotomy has played a central role in sociology since its inception. The approach based on this dichotomy enables the production of ideas on the axis of perceiving the village and the city as opposites of each other in terms of space, solidarity, population, education, and so on. It is essential to produce policies and regulations in a way to eliminate the unattractiveness of the countryside, which expresses the negativity of this idea. A strong economic structure that will make rural life attractive will be possible with a sustainable family business. In addition, we believe that the family business model, which is structured to sustain their lives in the countryside in order to prevent the concentration of immigrants in certain cities in our country, which is heavily exposed to foreign migration, will also be a solution to the immigrant problem. In this study, through the lens of the positive and negative contributions of aquaculture in the context of environmental, economic, social, and governance dimensions, the intersection of community resilience and sustainability of the food system is widely discussed. In this sense, (i) A comprehensive modeling of business networks, job performance, and family livelihood for aquaculture farmers operating in a difficult environment in Erzurum and developing theoretical knowledge and solution proposals and (ii) insight into the enrichment of family livelihoods and their mediating role by proposing new insights into how entrepreneurs can affect family livelihoods in a context where their families are at risk to meet their basic needs will be presented to provide contributions to the literature.

2. Conceptual framework

2.1 Aquaculture and family business

Family businesses have an important share and contribution in the world economy. It is possible to say that these enterprises have become a productive unit in terms of the added value they create and the functions they undertake. Therefore, their contribution to institutionalization and development by providing a socioeconomic field of activity in the countries where they are located is important. In addition, the fact that the process in family businesses is related to the behavior, performance, and interests of family members allows the issue to reach a social dimension beyond the economy. The obvious influence of family culture on business strategies requires thinking about the future of the business and the future of the family together ([4], p. 316). This allows for an overlap between business policy and family interests. Consequently, when the family is considered as an economic unit, each member of the family is expected to contribute to the general welfare, and it is understood that this situation is not limited to the family [5]. Family member or members are active in the management of family businesses. In the vast majority of these businesses, the key competitive advantage is the commitment and docility of the family member workforce to the organization and strong family ties. Employees are not seen as people with individual autonomy based on knowledge and skills but as people under the protection of the family or (at best) as foster children [6].

2.2 Migration phenomenon; definition, types, and explanations

Migration, which is defined as all displacements that take place at a distance that is meaningful and for a period of time that creates an impact [7], results in the interaction of groups and individuals with different cultures. Although this phenomenon, which is estimated to have started with human history, is defined as human mobility, it has reached many differences in terms of classification in the historical process. In this context, it is a population movement that changes the structure of society in economic, social, and cultural contexts with the process of changing space [8]. As can be understood from the definition, this phenomenon, which corresponds to spatial change, affects many factors related to society. At the very least, it is defined as moving away from the usual social structure and territory or economic system and opening the door to new living spaces [9]. It is also considered as the act of spending all or part of one's life for religious, economic, political, or social reasons in order to settle in another settlement unit. In addition, the phenomenon, which is pronounced as the redesign of the organization of space, has been one of the most discussed topics in the political, social, and academic community in recent years. In general terms, it is possible to divide migration into two groups according to the country border and according to its formation. According to the country border, migration is divided into internal and external migration. Internal migration is defined as the population movement of the citizens of a country within its borders.

It is explained as a type of migration in which the population of the country remains unchanged from one city to another city, from rural to urban or vice versa, or from one region to another region within the borders of the same country. Within internal migration, permanent migration is defined as individuals leaving their place of residence permanently and definitively and settling in another settlement unit. It is the voluntary or forced relocation of a person to another place for political, social, and economic reasons. It can be individual and voluntary for a number of reasons such as having better living conditions, or it can be forced by states to force their citizens to do so. In particular, natural disasters, terrorist incidents, and regional civil wars that occur unexpectedly cause forced permanent migration [10]. Another definition of internal migration is forced and voluntary migration. Voluntary migration is the relocation of people from one city to another city or region in line with their own wishes and expectations [11]. Migration between countries is conceived as an easier and more accessible relocation with globalization. However, the reading of foreign migration in recent years, especially in the axis of Turkey, reveals the fact that these migrations cause forced displacement.

2.3 Methodology

We based our study on a descriptive qualitative approach using a field study in Erzurum Province. The study was carried out in the center and districts of Erzurum between April 2022–September 2022 using the survey and interview method. During the research, 31 fish producers in the city were interviewed. Survey interviews were conducted in enterprises with different capacities in the selected area. Before the personal interview began, each of the farmers was given a brief introduction to the study and assured that all information would be kept confidential. Each question is clearly explained and systematically asked so that they understand it clearly. The time required for each interview was about an hour to an hour and a half. After collecting data from the field, the data were verified to eliminate errors and inconsistencies. The data were then carefully tabulated (**Table 1**).

Data were categorized and analyzed using MS-Excel software based on descriptive statistical analysis. All collected data were carefully accumulated and then processed to extract the findings of the study area.

		N	%
Gender	Male	30	96.8
	Female	1	3.2
Age	26–35	1	3.23
	36–45	9	29.03
	46–55	11	35.48
	56 and above	9	29.03
	not stating age	1	3.23
Marital status	Married	29	93.5
	Single	2	6.5
Education level	Primary school	6	19.4
	Secondary school	6	19.4
	High school	8	25.8
	Bachelor's degree	7	22.6
	Postgraduate	4	12.9
Place of birth	Erzincan	2	6.5
	Erzurum	25	80.7
	Rize	1	3.2
	Kars	2	6.4
	Yozgat	1	3.2
Farm Located County	Aziziye	3	9.7
	Yakutiye	2	6.5
	Tortum	4	12.9
	Uzundere	5	16.1
	İspir	9	29
	Kuzgun	2	6.5
	Oltu	3	9.7
	Pasinler	2	6.5
	Şenkaya	1	3.2
Monthly Income of the Business	10,000 TL (523 $) -50,000 TL (2615 $)	16	51.6
	50,001 TL (2615 $) -100,000 TL (5230$)	3	9.6
	100,001 TL (5230$) and above	2	6.5
	not stating income	10	32.3

Table 1.
Demographic characteristics of participants.

2.4 Results and discussion

Thirty-one fish farmers voluntarily participated in the research. In the 31 applied forms, there is no form that needs to be removed due to incomplete filling. *96.8%* of the participating pollsters in the research were male, and 3.2% were female. *It was observed that 3.23% were in the 26–35 age range, 29.03.3% were in the 36–45 age range, 35.48% were in the 46–55 age range, 29.03% were 56 years of age or older, and 3.23% did not indicate their age*. 93.5% of the participating pollsters in the research were married, and 6.5% were single; it is understood that 19.4% of them are primary school, 19.4% are middle school, 25.8% are high school graduates, 22.6% are undergraduate, and 12.9% are postgraduate. In addition, 80.7% of the participants were born in Erzurum, 6.5% in Erzincan, 6.4% in Kars, and 3.2% in Yozgat. It is understood that 9.7% of the enterprises are in the Aziziye district, 6.5% in the Yakutiye district, 12.99% in the Tortum district, 16.1% in the Uzundere district, 29% in the Ispir district, 6.5% in the Kuzgun region, 9.7% in the Oltu district, 6.5% in the Pasinler district, and 3.2% in the Şenkaya district. It is seen that 51.6% of the enterprises have a monthly income between 10,000 TL (523 $) and 50,000 TL (2615 $), 9.6% between 50,001 TL (2615 $) and 100,000 TL (5230$), and 6.5% of them have a monthly income of 100,001 TL (5230$) and above, and 32.3% do not specify their monthly income (**Table 2**).

Items	N	Min	Max	X	SD
FBSL8*	31	1	2	*1.23*	0.425
FBSL9	31	1	2	1.45	0.506
FBSL10*	31	1	2	*1.19*	0.402
FBSL11*	31	1	2	*1.03*	0.180
FBSL12*	31	1	2	*1.10*	0.301
FBSL13	31	1	2	1.45	0.506
FBSL14	31	1	2	1.29	0.461
FBSL15	31	1	2	*1.97*	0.180
FBSL16*	31	1	2	*1.32*	0.475
FBSL17	31	1	2	1.52	0.508
FBSL18*	31	1	2	*1.39*	0.495
FBSL19*	31	1	2	*1.39*	0.495
FBSL20*	31	1	2	*1.06*	0.250
FBSL21	31	1	2	1.42	0.502
FBSL22*	31	1	2	*1.16*	0.374
FBSL23*	31	1	2	*1.03*	0.180
FBSL24*	31	1	2	*1.13*	0.341

**FBSL8: ($\bar{x}$=1.23, sd = 0.425), FBSL10 ($\bar{x}$=1.19, sd = 0.402), FBSL11: ($\bar{x}$=1.03, sd = 0.180), FBSL12: ($\bar{x}$=1.10, sd = 0.301), FBSL16. ($\bar{x}$=1.32, sd = 0.475), FBSL18. ($\bar{x}$=1.39, sd = 0.495), FBSL19. ($\bar{x}$=1.39, sd = 0.495), FBSL20:($\bar{x}$=1.06, sd = 0.250), FBSL22:($\bar{x}$=1.16, sd = 0.374), FBSL23:($\bar{x}$=1.03, sd = 0.180), FBSL24:($\bar{x}$=1.13, sd = 0.341) was found. The item close to "no" was found to be FBSL15: ($\bar{x}$=1.97, sd = 0.180).*

Table 2.
Analysis results of family business and social life items (FBSL: family business and social life).

As a result of the analysis on family business and social life, it was seen that the average score was the closest to "Yes." The statistical values of these substances have significant explanations as indicated in **Table 2**.

Therefore, with the numerical values emerging in **Table 2**, it comes to the fore that the farmers make a living with the income they earn in this family business and that they think that the state incentive should be more and more comprehensive in order to continue family business. In this structure, where a significant part of the basic political decision-making bodies are members of a family, the assets necessary to maintain the family, management, *business, and operation* have a significant contribution to economic development and growth [12, 13]. In these structures, where family members are involved in the business and it is essential to carry out the business activity, another important issue stands out as a state incentive. In particular, despite the statements of the Ministry of Agriculture and Forestry that family enterprises engaged in animal production are included in the rural development support program, it is important that the fish farmers participating in the research need more state incentives and assistance in the scientific support points related to feed, fuel, infrastructure, and diseases. The demands and objections that have arisen in this direction show that they are closely related to the process of institutionalization of family business because institutionalization refers to the spontaneous execution of all functions and processes of an enterprise within certain rules and systematics [14]. Therefore, the conduct of business activities must depend on certain systematics. This situation contributes to the birth and development of business culture. In particular, it is important for the participants to be far from certain systematics and that this situation can only be eliminated with state plans and incentives. As a result of the analysis on family business and social life, this situation is seen more clearly when the items with the average score closest to "Yes" are examined because one of the most important pillars of the corporate transformation of family business and its management with a more systematic management is the collection of the budget of the enterprise and the planning of business activities. For the participants, one of the most important meanings of family business is the collection of the family budget in one hand. This situation is important for the continuity of the enterprise because with the budget planning, resource distribution is ensured in the enterprises, and in this sense, planning and organization performance are evaluated [15]. Only in this way can businesses plan budgets to build the future in a planned way, increase business profitability, and control management performance [16]. Collecting the budget in one hand and using it positively contributes to performance evaluation as well as providing control of activities [17].

The family as a social institution determines the basic subject and content of a scientific idea. Family business reveals the structure of a total family institution, its thinking and attitude. At the same time, the geographical location and climatic features of the family reveal the sociocultural and economic identity of the family. Another result of the study, in which the relationship between family business and social life was examined, is that regional characteristics determine the economic structure of the family and therefore contribute to the shaping of family business as an economic system. When **Table 2** is examined, the fact that the participants are under the influence of regional characteristics as an operator and that these characteristics affect the economic structure is revealed. While economic and social changes are influenced by existing family forms and intra-family relationships, on the other hand, these changes change family forms and intra-family relationships [18]. When the geographical locations where family businesses are located are examined,

DOI: http://dx.doi.org/10.5772/intechopen.1009287

it is understood that urbanization, industrialization, and trade activities have not developed significantly and create a process in which people living in the region have to resort to different economic ways to make a living. For this reason, when regional characteristics are taken into consideration, it is understood that people turn to family business and carry this to an economic activity area through aquaculture. In particular, the inadequacy of regional characteristics in terms of economic development reveals the fact that the participants turn to aquaculture. Other important parameters that support this situation are that the participants answered yes to the statements that "the geography we live in is effective in our family business" and "regional characteristics make a positive contribution to our economy." Obviously, when the climatic, commercial, and industrial characteristics of the nine regions where the family businesses are located (Aziziye, Yakutiye, Tortum, Kuzgun, Uzundere, Oltu, Pasinler, İspir, Şenkaya) are taken into consideration, it is understood that people turn to aquaculture, and this commercial activity contributes to their economies when geographical features are taken into consideration. This situation determines the entire field of activity of the family, which is the most important institution of the society, from economy to social life, cultural, and sociological, and shapes the family as a whole. One of the most important issues supporting this thought and analysis is again in the light of the data emerging in **Table 2**. The most important reason for family business operators to carry out this activity is the result of bringing good income. Therefore, family business contributes to the livelihood of the operators in a geography where regional characteristics are taken into consideration and economic development is not fully supported; however, it is understood that the state incentive needs to contribute more for the continuation of institutionalization and a systematic structure, and it is clearly stated that the most important reason for people to engage in this economic activity is to bring good income.

Family business is a model that is explained through business ownership and intergenerational transfer. And this model also becomes a means of the continuation of the family's togetherness, of spending time together [19]. As a result, businesses provide an opportunity for the development of a structure focused on long-term stability while allowing family members' commitment to the business to turn into a concrete form. Because in the system where the control belongs to a certain family, a content is important where the business is managed by the same family members and a family has had a say in the enterprise for several generations [20]. Family business is a company in which at least two generations work in the organization, where the purpose of the establishment of the enterprise is to provide for the family and prevent the disintegration of the inheritance, where the person who provides the family's livelihood manages the company, where family members are involved in a significant part of the management levels, or where family members are largely influential in making decisions [21]. In the section where the relationship between family business and social life is analyzed, the participants think that they can include the young people in the family in the business, that their children should do the same job, and that it is necessary to make it more attractive to live in the countryside in order for young people to continue family business. Therefore, there is a full belief that the enterprise will reach a more institutional structure through the transfer from generation to generation. Because family business is also the corporate structure that is initiated by an entrepreneur from the family and then the family is mostly involved in the business [22] blood ties are important at this stage, and the property belongs to the family. When the regional characteristics and family structures in which the study is carried out are taken into consideration, it is understood that the operators attach importance to blood ties especially for the continuation of the work. This situation has two different reasons and meanings. First, considering the geography where aquaculture is carried out, regional characteristics dominated by a rural life stand out, and it is expected that these regions away from the city will become more attractive and livable in order for young people to continue the same work. Secondly, the crown system is important for leaders in order for businesses established by a family member to take on a more institutional and systematic structure. People expect their children to do the same job and continue the family business. They state that both individual and state support are needed for this. It is understood that the participants who established the business and wanted the young people to do the same job in order to continue did not generally do this job as a father's profession. Therefore, the farmers who are not transferred from the institutional and generation to the generation and who turn to aquaculture individually and with the effect of geographical features, attach importance to the continuity of the business after them and hope that their family members will continue this business as a father's profession (**Table 3**).

In the relationship between family business, preventing migration from the countryside, and aquaculture, the participants expressed the idea that the form of family business prevents migration because it collects the budget in one hand and therefore can be a model for rural development and reverse migration to the countryside and that the continuation of young people in aquaculture can prevent migration from rural to urban. According to the pollsters, individual, legal, and state support are needed to prevent migration from the countryside and to ensure rural development. The state should establish policies for family business and increase its encouragement and support in order to prevent migration from the countryside. Young people should contribute to the institutional and established structure of family business and ensure its continuity and do this job. Support, incentives, and plans for family business should be legally guaranteed. It is expected that these steps will create a model. The model will try to stop rural-to-urban migration and even create reverse migration movements toward the countryside. This effort will not be limited

Items	N	Min	Max	X	SD
MIG25	31	1	2	1.26	0.445
MIG26	31	1	2	*1.84*	0.374
MIG27	31	1	2	*2.00*	0.00
MIG28	31	1	2	1.67	0.485
MIG29	31	1	2	*1.19*	0.402
MIG30	31	1	2	*1.03*	0.180
MIG31	31	1	2	*1.06*	0.250
MIG32	31	1	2	*1.03*	0.180
MIG33	31	1	2	*1.06*	0.250

**As a result of the analysis on migration, it was seen that the average score was the closest to 'Yes'. These substances; MIG29 ($\bar{x}$=1.19, sd = 0.402), MIG30 ($\bar{x}$=1.03, sd = 0.180), MIG31 ($\bar{x}$=1.06, sd = 0.250), MIG32 ($\bar{x}$=1.03, sd = 0.180), MIG33 ($\bar{x}$=1.06, sd = 0.250) was determined and the item close to "No" was obtained as MIG26 ($\bar{x}$=1.84, sd = 0.374), MIG27 ($\bar{x}$=2.00, sd = 0.00).*

Table 3.
Analysis results of ımmigration ıtems.

to internal migration and will allow migrants coming to our country to produce by settling them in the countryside, which will both prevent unregistered labor and prevent the exploitation of migrant labor. Therefore, family business will become an important and functional model in the relationship between migration and rural development. Because the phenomenon and dynamics of migration are shaped by many economic, cultural, and political determinants in the historical process in the sense of moving from one certain settlement unit to another, the necessity of using different concepts and assumptions periodically leads to the development of more than one theoretical approach that can explain migration movements. The formation of various perspectives on migration is directly proportional to the mobility revealed by the historical and social structure. In other words, migration in various societies and various historical processes is interpreted in relation to various contents and is the subject of social sciences. At this stage, migration as a displacement movement has an important function for rural development. In the historical process, migration has become an area of interest of societies for certain reasons. Demographic characteristics between societies, income differences between regions, economic crises, political instability, or civil wars are counted among these reasons. Along with its causes and consequences, it creates wide-ranging changes in individuals and societies. These changes should be read both through the migrating group and the indigenous people in the geography where they migrate because the process of social adaptation affects both segments. In this direction, adaptation or acceptance and adaptation or assimilation emerge as important areas that should be evaluated in a sociological context. One of the most important issues in the field where Turkey and migration topics are examined has been irregular migration. New and controversial concepts such as administrative surveillance, deportation, foreigners, migration flow, migration wave, citizenship grant, invitation to leave Turkey, and irregular migrant mass have emerged. Irregular migration, in this context, for target countries and for the country of origin, when covering persons who have come to their country illegally or who have come through legal means and have not left within the statutory exit period, covers people who cross the borders of the country by not following the necessary procedures when leaving the country. For a transit country, it is defined as people who enter the country legally or illegally to reach the destination country from the source countries and use this country as a transit route and leave the country's border. In other words, the concept allows a structure that has quite different meanings and should be supported by various facts at the point of definition. Considering the migration flows that Turkey has been subjected to since the 2010s, the social, economic, and political situations of the migrants are important. Therefore, it is important that the migrants who come in masses adapt to social life and are subject to a legal and legitimate system in economic life. This subordination may be made possible by a model that will be formed through family business. With this model, operators could employ irregular, cheap labor force and immigrants who are open to economic exploitation in the cities within the framework of aquaculture, allowing the rural population to turn into a mass contributing to the economy, because the country is witnessing waves of migration of millions of Syrians, Afghans, and citizens of various countries in an irregular way, and the demographic and economic structure is affected by this situation.

The nation-state and globalization affect various mechanisms at the point of giving and receiving immigration. Accordingly, nation-states have been decisive on those who enter and exit in terms of protecting their borders and integrity [23]. Stopping the masses that appear on the borders of nation-states and intend to enter the country has a responsibility to at least keep them under control and to stand with a certain will against the transformations that the incoming foreigners will cause in the social structure. Therefore, as a manifestation of their sovereignty, nations have to protect their borders, ensure that foreign elements are taken out of the country, and sanction against elements that are felt as threats. However, policies against all migration and restrictions on migration management do not resist the phenomenon of globalization very much because with globalization, there is an increase in the number of people leaving their country, and countries are deprived of the will to isolate themselves against this situation. As a result, supranational structures have shaped the policies of nation-states, while the security and control of borders have become more difficult. Efforts to evaluate migration on a global scale, such as the International Organization for Migration, aim to address migration beyond nations and to manage mobility within the plan and project. In this context, the agenda of international politics is similar to cooperation and general approaches and migration management. In this direction, the situation that migration cannot be managed in a single center brings about the production of common policies because migration, technical development, and increase in transportation and communication opportunities facilitate relocations, and mobility gains speed. In addition, the enforcement power of multinational corporations, international businesses, and money markets and the displacement of more people caused by the tourism sector have put forms of migration into new forms [24]. Migration has now become globalized, accelerated, and evolved into new forms. In addition, when migration reaches an unplanned and uncontrollable dimension, it has become politicized, becoming more associated with security policies. In a sense, globalization also causes an increase in the volume of international migration by allowing diversity in the reasons for migration. As a result, the effort to protect the borders and integrity of nation-states has had to cope with developments on a global scale. When all these developments and results are taken into consideration, it comes to the fore that the issue of migration is an important social and political issue for Turkey and that it should be handled with various dimensions. Family business has many meanings for the operators participating in the study. The first creates a budget as an economic activity in the countryside and prevents the migration of operators and their families because it allows the budget to be collected in one hand. Secondly, both rural migration and foreign migration to the country are an important sociological issue in Turkey. To contribute to the significant resolution of this issue, family business can serve as a model. Supporting family business with legal and state incentives will allow regular/irregular migrants to be both registered and operators to maintain the existing system. Therefore, this model will open the door for migrants coming to our country to make production by settling them in the countryside, to social adaptation, to prevent cheap labor force, and to enable migrants to live in more humane conditions and without being exploited economically. The fish farmers believe that their own children should continue to run the family and contribute to rural development with the argument that it is important for generations to stay in the countryside. Another piece of data that supports these issues is the answer that the participants did not think about migrating when **Table 3** was examined. Without exception, no participant is willing to emigrate. Participants who express their desire to live in the regions where the business is located have the aim and intention of staying in the existing settlements with their families. They emphasize the importance of migration to the countryside, not migration from the countryside.

2.5 Exploratory factor analysis

The main purpose of factor analysis is to reduce or summarize to a smaller number of fundamental dimensions to facilitate the understanding and interpretation of patterns between the large numbers of variables that are thought to be related between them. In other words, it is a method of reducing to a dimension such as basic component analysis and destroying the dependency structure [25, 26]. As it is understood, there are two main purposes of factor analysis [27]. These are to reduce the number of variables and to create some new structures by taking advantage of the relationships between variables. Explanatory factor analysis is a technique that is used to derive a smaller number of variables ($k < p$) and new independent variables (factors) from a number of p variables related to each other by using the covariance or correlation matrix of the data [28]. The results of the factor analysis of the scales used in the research (for family business, economy and policies, family budget, attitude of family members and the impact of the region) are summarized in **Table 4**.

As a result of the factor analysis, it is seen that five sub-factors related to family business and social life scale explain 81.005% of the total variance (items 10–13 are removed from the scale). The operators participating in the study on the family

Family business and social life (**α = 633, total explained variance 81.005%, KMO = 0.028**)	
Family Business	
s15	−0.932
s23	0.932
s20	0.674
Economics and Politics	
s18	0.866
s17	0.814
s16	0.778
s9	−0.471
Family Budget	
s11	0.925
s24	0.647
s8	0.557
Attitude of Family Members	
s21	0.796
s14	0.740
s19	0.684
Influence of the Region	
s12	0.792
s22	0.585

It was seen that they explained 81.005% of the sample. Items 10 and 13 have been removed.

Table 4.
Results of exploratory factor analysis of the scales used in the research.

	Value	df	Asymp. Sig. (2-sided)
Your gender	5.373	1	.020
Your marital status	.411	1	.521
Your age	.455	3	.929
Your Education Status	6.183	4	.186
Income From Family Business Provides Our Livelihood	.023	1	.880
Family Business Prevents Our Migration	2.074	1	.150
I Consider Doing The Same Job Where We Migrated	1.125	1	.289

Table 5.
Chi-square test table- migration behavior and the relationship between variables consider migration.

business sub-factor state that family business is not the father's profession, but the geography they are located in is an important reason for them to turn to family business. In the light of factor analysis data especially, the participants revealed that they started aquaculture because the geography they lived in was deprived of industrialization and urbanization opportunities and that they did this in the form of a family business. In this case, family business needs to become permanent by including other generations because it is considered an important step to make living in the countryside attractive for young people to continue their family business. The participants think that young people should continue the businesses they have established due to geographical/regional characteristics that are not the father's profession and that this will both make rural living attractive and support rural development. In the sub-factor of economy and policies, the participants who stated that they do family business because it brings good income state that they are in contact with the relevant sector components but are not aware of the agricultural policies related to family business.

This situation shows that the fish farmers are not aware of the current developments and that they are inadequate in terms of internet and information exchange, especially considering the age range of the participants. The family budget sub-factor shows that the participants make a living with the income they earn from family business that the economic activity in question gives them the opportunity to collect the budget in one hand, and therefore, the family members are reluctant to migrate. Considering the regional features, especially the presence of dams, have allowed people to establish a family business through aquaculture, and the economic situation arising from this business have affected many institutional and social structures. The structure of the family, its economic situation, the shaping of the will to migrate, regional characteristics, and geographical inadequacies have a direct connection and relationship with each other. In this geography, where there are no different business lines and opportunities, people started aquaculture due to dams, designed it as a family business, started to earn a good income, and migration was prevented by collecting the income in one hand.

The attitude of family members is also important in terms of the existence, continuity, and institutionalization of the family business. The sub-factor entitled attitude of family members suggests that participants are supported and welcomed by the household to continue the family business. Stating that there is a cooperation on aquaculture in the family, the participants state that the young generation contributes significantly to the business and that there is a blood connection-oriented coordination and sharing. In addition, with the acceptance that family business can be a model for rural development, the attitude of hope that the new generation will continue to run family business is important. This situation is in parallel with the result of the sub-factor of the impact of the region because in this sub-factor, when the operators take into account the regional characteristics, they think that their children should do the same job because these participants show resistance at the point of migration and accept that aquaculture is the best economic initiative for their children to make considering the current conditions that they are in.

2.6 Chi-square test

When the chi-square test in **Table 5** is taken into consideration, there are significant and meaningless relationships between migration behavior and various parameters. First of all, men constitute the economic power in the enterprises. When the geographical and cultural region in question is taken into consideration, the existence of a patriarchal structure and system draws attention (**Tables 5** and **6**).

Men are the determinants and owners of economic power, which is accepted as the power to approach economic necessity with a certain distance. Therefore, the power that holds the means of production has an economic, social, and cultural influence and inclusive power. Because patriarchy is the area where men determine their rule and power over women, it has the feature of a "system." However, patriarchy is not a fixed system but a supra-system 'structure' and a network of relations between representation and subject. Because the man is surrounded by the patterns of being a man, "acting like a man," and so on from the moment he is born, he has to constantly understand the process and fulfill his responsibilities. In this process, unlike the woman, there are stages of separation and steps to be taken in the maturation process of the man. In this structure, where the most important thing is to learn the masculine language, the struggle of the man with different models and his inner impulses, his relations with other people are seen as indispensable stages to prove his own masculinity. Therefore, being a man, recognizing the limits to the extent required by masculine language, requires structuring within a never-ending mechanism of expectation and control [29]. It is understood that the operators in question have turned to family business in order to hold economic power due to regional characteristics and the fact that they are men. Therefore, the economic power of enterprises is shaped in exchange for the performance and "labor" that the culture and social structure expect

Chi-square tests					
	Value	**df**	**Asymp. Sig. (2-sided)**	**Exact Sig. (2-sided)**	**Exact Sig. (1-sided)**
Pearson Chi-Square	6.147[a]	1	.013		
Continuity Correction[b]	2.702	1	.100		
Likelihood Ratio	5.834	1	.016		
Fisher's Exact Test				.060	.060
Linear-by-Linear Association	5.948	1	.015		
N of Valid Cases	31				

Table 6.
Chi-square test table-family business and migration behavior.

from men. Considering the necessity of a continuous construction of men in the patriarchal structure, it becomes possible to say that the men who established family business have reached various forms of power, starting from the economy. As a result, in the table where the relationship between the variables is tested, it is seen that men hold economic power and can direct migration behavior thanks to this *power*.

When the relationship between migration behavior and variables is examined, there is a significant relationship between gender and migration thinking behavior. It is understood that male business owners do not think about immigration; the most important parameter that prevents them from migrating is having a family business. Therefore, it is clear that the people who hold economic power determine many sociocultural and social perspectives of the family. It is understood that the family as a whole does not have the desire to migrate due to the male business owners who resist at the point of migration. When the relationship between the same variables is examined, the fact that people are married or single as well as age and education status are not the determining factors in migration behavior. In other words, in the geography where the gender variable shapes the migration behavior due to the fact that the business owners are men, people (young or old, married or single) express that they do not think about immigration. Therefore, there is a significant relationship between educational status, marital status, and age and variables of thinking about immigration. Participants do not consider migrating due to reasons such as regional characteristics, income, and efforts to support rural development in theory and practice. The same participants are not willing to migrate because they have a family business and the income they earn from this business, and they also think that the region where they live should become more attractive and reverse migration should begin. He states that the most important task at this stage belongs to future generations, state officials, and various policies.

When the table examining family business and migration behavior is taken into consideration, it has been analyzed that family business prevents migration in the current situation, but if young people continue to operate their family businesses, migration from the countryside can be prevented. As a result, there is a significant relationship between the variables "family business prevents us from migrating" and "the continuation of family business by young people can prevent migration from the countryside." And while this relationship clearly reveals the obstacles to the migration of participants with family businesses, it also presents the recipe for preventing migration from the countryside as the continuation of this work by young people.

3. Result and discussion

As it is known, continuity in agricultural production is very important in terms of continuity in production by not fragmenting the land and keeping family members together. For this, it is necessary not to fragment the land through inheritance and to collect the budget in one hand. For the participants in our study, one of the most important features of family business is that it allows the family budget to be collected in one hand. This is important for the continuity of the business. A large proportion of the operators participating in the research stated that family businesses collect the budget in one hand. Collecting the budget in one hand and using it in a way that satisfies family members provides control of activities and contributes to performance evaluation. Just as institutions in society have functions, businesses also have performance and functional roles. Among the functions of family business, we can count the ability to continue its existence, to protect the family unity, to collect the budget in one hand, to survive economically, and to make savings. In modern societies, the problem of function has become a problem of performance.

Public influence is gained when performance roles are fulfilled effectively and efficiently. As a result of the interviews we conducted with the participants, it is an important conclusion that performance roles are fulfilled effectively thanks to the functions that arise with the continuity of aquaculture family business, and as a result, trust and public impact can be created. It is seen that regional characteristics are important in the participants' orientation toward family business and that they realize an economic activity area through aquaculture. In particular, the inadequacy of regional characteristics in terms of economic development is an important factor for the participants to turn to aquaculture. Other important parameters that support this situation are that the respondents mostly answered "yes" to the statements "the geography we live in is effective in family business" and "regional characteristics contribute positively to our economy." Obviously, considering the climatic, commercial, and industrial characteristics of the nine regions (Aziziye, Yakutiye, Tortum, Kuzgun, Uzundere, Oltu, Pasinler, İspir, Şenkaya) where the family businesses are located, it is understood that people are oriented toward aquaculture and that this commercial activity is economically satisfactory considering the geographical characteristics.

Therefore, when the regional characteristics of the family businesses are taken into account and in a geography where economic development is not fully supported, the fact that it brings good economic income makes it necessary for the state to contribute more in terms of showing that it is feasible for young people, even if the countryside is repulsive. In the section of the study where the relationship between family business and social life was analyzed, the participants stated that they were able to involve the young people in the family in the business and that they wanted their children to do this business, which is one of the important pillars of the institutionalization process; the other is state support, and the third is to make structural changes that will enable people to turn to family business in rural areas. Thus, institutionalization and a systematic functioning will be ensured by transferring from generation to generation. Since family business refers to an institutional structure initiated by an entrepreneur from the family and then the family is mostly involved in the business, blood ties are important, and ownership belongs to the family. It was emphasized by the participants that blood ties were important for the continuation of the business. The stagnation of rural life in aquaculture is an important factor that breaks the desire of young people to continue this business. Although people want their children to do the same job and continue the family business, it is certain that there is a need for socialization tools that will make life more livable for young people. The fact that the operators state that they do not do this profession as a father's profession but that their children will take it over from them is an important indicator that family business will continue with two generations practicing this profession. Each participant stated that individual efforts are insufficient and that government incentives and support are essential in the argument that family business management, which is included in the scope of the study, will be a model for preventing rural migration and reverse migration. One of the most important reflections of the phenomenon of migration, which is accepted to have started with the history of humanity, is international migration and illegal migration. It is a reality that there are both types of migration to our country, and this has severe consequences. The problem of international migration has been on the agenda especially since the last quarter of the twentieth century. Due to reasons such as political instability, human rights violations, oppressive regimes, civil wars, lack of labor demand, economic difficulties, inadequate geographical conditions, and life safety, people migrate to countries where they hope to live better. Turkey is one of the leading countries exposed to intensive migration both as a destination and source country. With the family business model, our study proposes that it will stop rural–urban migration and even create reverse migration movements toward the countryside, as well as allowing migrants coming to our country to produce by placing them in the countryside, thus preventing unregistered labor and preventing the exploitation of migrant labor. In today's Turkey, it is essential that the policies produced regarding the social, economic, and political situation of immigrants should provide permanent solutions in the long term. The adaptation of migrants arriving in masses to social life and their presence in economic life requires them to be subject to a legal and legitimate system. This subordination can be made possible through a model that will be formed through family business. With this model, operators will not only prevent irregular, cheap labor and economically exploitative conditions in the cities but also allow the rural population to return to a mass that contributes to the economy. Because we are witnessing irregular migration waves of millions of Syrians, Afghans and citizens of various countries, demographic and economic structures are affected by this situation. In this study, we are interested in the applicability of the family business model as a policy to prevent migration by underlining that migration from the countryside as internal migration and the problems arising as a result of this migration and the solutions produced to these problems do not prevent migration, and in addition to this, immigrants coming to our country intensively with foreign migration cause new problems, and the importance of the migration policy produced for this in the national and international arena is increasing. In this sense, there are various reasons why the number of male operators is high in the profile of the breeders who voluntarily participate in the research. It is the fact that the work done is a family profession and that the person who continues the family profession in our society is generally male and the number of male operators is closely related. In addition, both the harsh geographical conditions of the region where the enterprise is located and the extension of the patriarchal cultural remains are also effective in the high number of male operators. The participants stated that they are doing this job because the income they earn in family business is satisfactory and that they are struggling economically in the context of yesterday and today and that the state incentive should be more and more comprehensive in order for young people to continue their business after them. Despite the support of the Ministry of Agriculture and Forestry, they have often stated that they need more state incentives and assistance in terms of feed, fuel, infrastructure, and scientific support on diseases.

4. Conclusion

Although our study focused on Erzurum as a pilot region, we argue that our findings have the potential to be generalized to other contexts. This is because it is based on the recommendations of existing transition research to go beyond individual case studies. In our study, we carefully established a national example of countries with heterogeneous industrial transition models in terms of both direction and magnitude of various sustainability actions/performance and different socio-technical levels of development.

DOI: http://dx.doi.org/10.5772/intechopen.1009287

Based on this example, we believe that developed economies that share certain socioeconomic characteristics of Turkey can find our findings applicable to their own sustainability situations.

Data availability statement

The data that support the findings of this study are available from the corresponding author upon reasonable request.

Author details

Nuray Karaca[1*], Erdi Aksakal[1], Cengiz Kılıc[2], Arzu Uçar[3], Veysel Parlak[4], Muhammed Atamanalp[3] and Gonca Alak[5*]

1 Department of Sociology, Faculty of Literature, Ataturk University, Erzurum, Turkey

2 Department of Sociology, Faculty of Open and Distance Education, Ataturk University, Erzurum, Turkey

3 Department of Aquaculture, Faculty of Fisheries, Ataturk University, Erzurum, Turkey

4 Department of Basic Sciences, Faculty of Fisheries, Ataturk University, Erzurum, Turkey

5 Department of Seafood Processing Technology, Faculty of Fisheries, Ataturk University, Erzurum, Turkey

*Address all correspondence to: nkaraca@atauni.edu.tr and galak@atauni.edu.tr

References

[1] Carrera-Quintana SC, Gentile P, Girón-Hernández J. An overview on the aquaculture development in Colombia: Current status, opportunities and challenges. Aquaculture. 2022;**561**:738583, 1-2

[2] Taylor JG. The emergence of the articulation of production modes and ıts effect on the structure of third world formations. In: Keskinok HÇ, Ersoy M, editors. On the Articulation of Production Modes. Ankara: Individual and Society Publications; 1984. pp. 125-178

[3] Karaca N. On the consequences of modernization in Turkiye. Erzincan University Journal of Social Sciences Institute. 2020;**2020-13**(2):117-128

[4] Sevinç I. Problems facing family businesses in the growth process: An application in family businesses operating in Konya, Atatürk Unv. Journal of Economics and Administrative Sciences. 2005;**19**(2):315-331

[5] Güney S. Success conditions in family businesses in the process of transfer to the second generation. Journal of Entrepreneurship and Development. 2007;**2**(1):103-126

[6] Asunakutlu T, Avcı U. A research on the relationship between nepotism perception and job satisfaction in family businesses. Journal of the Faculty of Economics and Administrative Sciences of Süleyman Demirel University. 2010;**15**(2):93-109

[7] Erder S. "Ghetto" and Turks in Welfare Society. İstanbul: Technographic Printing; 1986

[8] Özer İ. Urbanization, Urbanization and Urban Change. Bursa: Ekin Bookstore; 2004

[9] Toros A. Migration in Problem Areas. Ankara: Global Strategy Institute; 2008

[10] Koçak Y, Terzi E. The phenomenon of migration in Turkiye, the effects of ımmigrants on cities and solution suggestions. Journal of the Faculty of Economics and Administrative Sciences of Kafkas University. 2012;**3**(3):163-184

[11] Özkalp E. Introduction to Sociology. Eskişehir, Turkey: Eskisehir Anadolu University Press; 1990

[12] Olson PD. The impact of the family and the business on family business sustainability. Journal of Business Venturing. 2003;**18**:639-666

[13] Neubauer H. The dynamics of succession in family business in western countries. Family Business Review. 2003;**14**:4, ABI/INFORM Global, 269

[14] Yazıcıoğlu İ, Koç H. A comparative research on the determination of ınstitutionalization levels of family businesses. Selçuk University Journal of Social Sciences Institute. 2009;**21**:497-507

[15] Fisher JG, Maines LA, Peffer SA, Sprinkle GB. Using budgets for performance evaluation: Effects of resource allocation and horizontal information asymmetry on budget proposals, budget slack, abd performance. The Accounting Review. 2002;**77**(4):847-863

[16] Uysal G. Budget process and applications in providing organizational source and target relationship: Budget planning, execution and control. Journal of Accounting and Finance. 2008;**38**:113-122

[17] Kung FH, Huang C, Li CCL. An examination of the relationships among

budget emphasis, budget planning models and performance. Management Decision. 2013;**51**(1):120-140

[18] Ökten S. The effects of socio-cultural and structural characteristics of the GAP region on the family structure. Family and Society. 2006;**3**(9):23-34. DOI: 10.21560/spcd.09353. January-February-March 2006. ISSN: 1303-0256

[19] Jaffe DT. Working with the Ones you Love: Conflict Resolution and Problem Solving Strategies for a Successful Family Business. Berkeley, CA: Conari Press; 1990

[20] Neubauer F, Lank A. The Family Business: Its Governance for Sustainability. Houndmills, London: MacMillan Press Ltd., Palgrave Macmillan; 1998. pp 65-67

[21] Ateş Ö. Family Companies: Change and Continuity. Ankara: Ankara Chamber of Industry Publication; 2005

[22] Findikçi İ. Family Firm in Turkey: The Struggle for Change. Family Business, Winter; 2005. pp. 11-13

[23] Sunata U. Your Diasporic identity, please! Chinese diaspora in the USA. In: Ihlamur-Öner SG, Öner NA, editors. Migration in the Era of Globalization. İstanbul: Communication Publications; 2012. pp. 459-482

[24] Aksoy Z. International migration and ıntercultural communication. International Journal of Social Research. 2012;**5**(20):294-295

[25] Korkmaz A. Finalization of criminal judgment process with factor analysis and non-parametric techniques, reviewing the process [Doctoral thesis]. Ankara: Hacettepe University; 2000

[26] Tatlidil H. Applied Multivariate Statistical Analysis. Ankara, Turkey: Cem Web Ofset Ltd.Sti; 1996

[27] Özdamar K. Paket programlar ile istatistiksel veri analizi. (4. Baskı). Eskişehir: Kaan Yayınları; 2002

[28] Doğan N, Başokçu TO. Comparison of factor analysis applied for statistical attitude scale and stepwise clustering analysis results. Journal of Measurement and Evaluation in Education and Psychology, Winter 2010. 2010;**1**(2):65-71

[29] Türker Y. Power and man. Society and Science. 2004;**101**:8-10

Chapter 2

Aquaculture in Africa: Production Systems, Challenges, and Opportunities

Shadrack Kwadwo Amponsah, Divine Dogbeda Azumah and Seth Koranteng Agyakwah

Abstract

Aquaculture in Africa presents a critical avenue for addressing food security challenges, economic development, and environmental sustainability. This chapter explores the continent's aquaculture production systems, including traditional methods such as earthen ponds, tanks, and cages, and advanced technologies such as Recirculating Aquaculture Systems (RAS). While aquaculture contributes significantly to dietary protein and economic livelihoods, its expansion faces challenges such as limited access to quality feed, water scarcity, and financial barriers. By analyzing the barriers and identifying opportunities for growth, this chapter underscores the potential of integrated systems and innovative practices to promote sustainable aquaculture development. Case studies, such as Ghana's Wontesty Ventures and Kenya's Victory Farms, illustrate successful models of sustainable aquaculture contributing to local economies and food security. Recommendations for policy support and strategic investments are also presented to encourage public-private partnerships and capacity-building initiatives. The findings highlight aquaculture's transformative role in ensuring food and nutrition security, promoting economic resilience, and safeguarding aquatic ecosystems for future generations.

Keywords: food security, sustainable aquaculture, policy support, innovation, RAS

1. Introduction

The world's population is expanding, increasing the demand for food, especially animal protein [1]. Fish is an important source of protein, vitamins (A, B12, D, and E), calcium, iodine, iron, selenium, and zinc, often lacking in most human diets [2]. In Africa, fish is a crucial dietary component, contributing to 25–40% of the protein needs for over one million people living around the African Great Lakes [3]. Per capita fish consumption in Ghana, for instance, is estimated at 26 kg, surpassing the global average of 20 kg and the African average of 10 kg [2]. This reliance highlights the importance of the fisheries sector in supporting livelihoods and economic growth across the continent.

Aquaculture offers a potential solution to the growing demand for fish protein, promoting food security while potentially alleviating pressure on wild fish stocks [4]. While aquaculture is expanding globally, its contribution in Africa remains relatively small, despite recent growth [5, 6]. Africa accounted for 13.1 million tonnes of fisheries and aquaculture production, 6 percent of the world total [7]. The vast majority of African aquaculture (99%) relies on inland freshwater systems, mainly for the culture of indigenous species such as tilapia and African catfish [8].

To meet the growing demand for aquatic protein sustainably, a comprehensive understanding of African aquaculture production systems, challenges, and opportunities is essential. This chapter will review these aspects of aquaculture in Africa, with a focus on the various aquaculture systems prevalent across the continent, encompassing earthen ponds, cage culture, and integrated aquaculture-agriculture systems. Also, examine key obstacles impeding the growth of aquaculture in Africa. This will involve examining factors such as inadequate infrastructure, limited access to quality fish feed, financial constraints, and the impact of climate change, particularly flooding and droughts. Lastly, explore the diverse opportunities available to foster sustainable aquaculture development in Africa. This encompasses the promotion of technological advancements, including the integration of aquaculture with other agricultural systems.

This study will highlight successful initiatives and best practices, while also providing recommendations for policy interventions and investments to promote sustainable and resilient aquaculture development in Africa. This analysis aims to contribute to strategies for ensuring food and nutrition security for the continent's expanding population, promoting economic growth, and safeguarding aquatic ecosystems for future generations.

1.1 Aquaculture production systems in Africa

Aquaculture has emerged as a crucial contributor to food security, economic development, and sustainable livelihoods across Africa [9]. With the continent's growing population and increasing demand for fish, aquaculture production systems have become vital to ensuring adequate fish supply [8]. Africa is rich in aquatic resources, but there is a need to balance between traditional systems and the adoption of emerging technologies to enhance fish production [10]. This section explores the primary aquaculture production systems in Africa, focusing on traditional methods such as earthen ponds, tanks, and cages, alongside emerging technologies such as the Recirculating Aquaculture System (RAS).

1.2 Traditional aquaculture systems in Africa

Traditional aquaculture systems in Africa have been practiced for decades, primarily relying on local knowledge, minimal inputs, and simple infrastructures [11]. These systems are mainly based on the use of earthen ponds, tanks, and cages, which have been instrumental in meeting local fish demands serving as source of income for rural households.

1.2.1 Earthen ponds

Earthen ponds are a popular choice for aquaculture, particularly in Africa, due to their low construction cost and ability to create a natural environment for fish [12]. These ponds are typically constructed using natural soil and can vary significantly in size—from small backyard ponds to larger commercial operations spanning several hectares [13]. There are two main types of earthen pond systems: extensive and semi-intensive. Extensive systems rely on natural productivity with minimal management; fish are primarily fed on natural food sources such as algae and insects [11]. This approach is cost-effective for small-scale farmers but yields lower production rates—averaging around 250 kg/ha/year [14]. In contrast, semi-intensive systems involve some level of feed input and management practices to enhance fish growth while still benefiting from natural productivity [15]. The case of Ghana illustrates the effectiveness of earthen ponds in promoting local aquaculture. Farmers have successfully utilized these systems to cultivate tilapia and catfish, contributing significantly to local diets and economies (**Figure 1**) [16].

1.2.2 Tanks

Tank-based aquaculture is gaining traction as a more controlled environment for fish farming [17]. Tanks can be constructed from various materials including concrete, fiberglass, or plastic [18]. This method allows for better control over water quality and temperature, which can lead to improved fish growth rates [19]. They are frequently employed for fry rearing, providing a controlled environment for the delicate early life stages of fish. Concrete and fiber tanks are also utilized in hatcheries with recirculating troughs for incubating fertilized eggs [20]. However, tank systems come with challenges. They often require higher initial capital investments compared to earthen ponds and demand more technical knowledge for effective management [21]. In Nigeria, for example, recirculating aquaculture systems (RAS) have been adopted within tank-based systems for intensive catfish production, especially in urban areas with limited land availability to maximize efficiency and sustainability [22]. These systems enable farmers to recycle water while maintaining optimal conditions for fish health (**Figure 2**).

Figure 1.
A view of an earthen pond used for fish farming, showcasing its natural design and large water retention capacity for aquaculture purposes.

Figure 2.
Tank culture system highlighting the use of artificial enclosures for controlled aquaculture, suitable for efficient fish farming and water management.

1.2.3 Cages

Cage aquaculture involves raising fish in floating enclosures placed in natural water bodies such as lakes or rivers. This method is particularly popular in regions with abundant water resources. Cages can be constructed from various materials such as nets and drums designed to allow water flow while keeping fish contained [23]. The number of cages in a given water body is limited by factors such as depth, water current, and wind velocity, all affecting water circulation [12]. Cage aquaculture relies on

Figure 3.
Cage tank system illustrating floating or submerged enclosures in open water bodies, designed for efficient space utilization and high-density fish culture.

the natural water body's quality, making factors such as temperature, salinity, oxygen levels, and pollution crucial considerations [24]. Cage systems allow for higher fish production in the same water body compared to capture fisheries. For instance, a small water body's natural productivity might be limited to 300 kg/ha, but using it for fed cages could increase production to at least 3 tons per hectare [12]. In Ghana, cage systems account for 90% of farmed fish production in Ghana, making it the leading aquaculture business industry [25]. The majority of cage farms are situated on Lake Volta, particularly in the Eastern Region, specifically the Asuogyaman District [26]. The lake's suitability for cage farming, coupled with government support, has contributed to this concentration (**Figure 3**).

2. Integrated aquaculture systems

Integrated aquaculture systems combine fish farming with other agricultural practices to create sustainable farming environments [11]. Common forms include rice-fish farming and aquaponics. In rice-fish farming, fish help control pests while benefiting from nutrients released by rice plants [27]. Aquaponics involves growing fish alongside plants in a symbiotic environment where fish waste provides nutrients for plants [28]. These integrated systems enhance productivity and resource efficiency but may require more complex management strategies. They represent a promising avenue for improving food security while promoting sustainable agricultural practices.

2.1 Emerging technologies in aquaculture: The role of recirculating aquaculture systems (RAS)

As the demand for seafood continues to rise globally, sustainable aquaculture practices have become essential to meet this need while minimizing environmental impacts. Among the innovative solutions emerging in this field, recirculating aquaculture systems (RAS) stand out as a transformative approach. RAS technology not only enhances water efficiency but also offers adaptability and reduced environmental impact, making it a vital component of modern aquaculture.

2.2 Overview of recirculating aquaculture systems (RAS)

Recirculating aquaculture systems are designed to treat and recycle water within a closed-loop system. Unlike traditional aquaculture systems that rely on a constant flow of fresh water, RAS significantly reduces water consumption by filtering and reusing water [29]. RAS technology was initially developed in Japan in the 1960s, motivated by a shortage of water for fish rearing [30]. This method can achieve water savings of up to 90% compared to conventional open systems, making it an attractive option for aquaculture in areas where freshwater resources are limited [31, 32].

3. Benefits of RAS

3.1 Water efficiency

One of the most significant advantages of RAS is its exceptional water efficiency. By continuously recirculating and treating water, RAS minimizes the need for large volumes of freshwater [33]. It also conserves water by treating and reusing a large portion of the water, requiring only minimal replenishment [34]. This is particularly crucial in regions facing water scarcity or where aquaculture operations are situated in urban environments where access to natural water bodies is limited [5]. The closed nature of RAS also helps prevent nutrient-rich effluents from polluting surrounding ecosystems, thereby mitigating environmental degradation associated with the traditional aquaculture practices [29].

3.2 Adaptability

RAS systems are highly adaptable and can be implemented in various settings, including urban areas and land-scarce environments [35]. This flexibility allows for year-round production independent of external climatic conditions, ensuring a consistent supply of high-quality seafood [36]. Furthermore, RAS can accommodate a broad range of species, including those with specific temperature requirements, making it suitable for diverse aquaculture operations [33]. RAS can be designed and tailored to meet the specific environmental and water quality requirements of different species, allowing for the culture of a diverse range of fish, crustaceans, mollusks, and even aquatic plants [37].

3.3 Reduced environmental impact

The environmental benefits of RAS extend beyond water conservation. RAS has been designed to drastically reduce water usage compared to traditional flow-through systems [38]. Water is treated and recirculated within the system, resulting in water replacement rates as low as 1% per day [31]. Advanced RAS designs incorporate denitrification reactors to further reduce nitrate levels, minimizing the risk of nutrient pollution in surrounding water bodies [39]. RAS offer the potential for nutrient recycling, further minimizing environmental impact and contributing to a circular economy approach [37]. Research is exploring the use of RAS waste for biogas production, offering a sustainable energy source and further reducing waste [40]. The enclosed and controlled environment of RAS minimizes the risk of disease transmission to wild fish populations and the surrounding ecosystem [31]. Aich et al. [39] stated that RAS offer better control over water quality parameters, reducing stress on fish and enhancing their immune systems, making them less susceptible to diseases. The limited water exchange in RAS restricts the entry of pathogens from external sources.

3.4 Challenges facing aquaculture in Africa

Aquaculture in Africa presents a promising avenue for enhancing food security, economic development, and employment opportunities [41]. Despite the promising potential, several challenges need to be addressed to achieve sustainable aquaculture development in Africa.

4. Resource constraints in African aquaculture

4.1 Limited water availability

Water is a fundamental resource for aquaculture, and its scarcity poses a major challenge across many African regions [42]. Water scarcity in many regions of Africa restricts the establishment and expansion of aquaculture operations [43]. The continent's freshwater resources are unevenly distributed, with some areas experiencing severe droughts while others face flooding [44]. For instance, countries such as Nigeria and Uganda have abundant water bodies suitable for aquaculture, yet the overall management of these resources is often inadequate due to pollution and competing demands from agriculture and domestic use [8]. Furthermore, the degradation of water quality due to urbanization and industrial activities exacerbates the situation [8]. As reported, many water bodies around urban centers are heavily polluted, making them unsuitable for aquaculture practices.

4.2 Feed constraints

Fish feed is the primary source of nutrients in aquaculture systems, and its quality and availability directly influence fish growth and productivity [45]. The production of high-quality aquaculture feed is another critical constraint affecting growth in this sector. Kaleem and Bio Singou Sabi [11] point out that fish farming is bedeviled with constraints such as high cost of fish feed and Shaalan et al. [28] highlight that fish feed contributes to 75–85% of the running costs in fish production, making it a substantial financial burden for farmers. This is because most farmers in Africa rely heavily on imported feed ingredients and fish feeds from European countries, making fish farming expensive [46]. This reliance on imports makes the aquaculture sector vulnerable to global price fluctuations and supply chain disruptions. In Uganda, the lack of government regulation on fish feed quality assurance has led to inconsistencies in the standard of locally manufactured fish feed [8]. This uncertainty about feed quality poses a significant risk to fish farmers, as they cannot guarantee the nutritional value of the feed they are using, potentially leading to reduced growth rates and increased disease susceptibility in their fish.

4.3 Land availability

Land availability for aquaculture development is also limited in many regions of Africa. The most prominent competition for land suitable for aquaculture comes from agriculture. In Egypt, South Africa, and Nigeria, competition for land use between agriculture, urban development, and other industries has restricted the potential sites for aquaculture farms [8, 47]. In some cases, existing agricultural practices may conflict with aquaculture operations, leading to resource mismanagement and environmental degradation [48]. Moreover, land tenure issues can deter investment in aquaculture as potential investors may be reluctant to engage in long-term projects without secure land rights [16].

4.4 Financial barriers

4.4.1 High initial costs

One of the primary financial challenges facing aquaculture in Africa is the high initial investment required for implementing RAS technology [49]. RAS facilities are designed to provide a controlled environment for fish farming, utilizing a closed-loop system that recirculates water, thereby minimizing waste and optimizing resource use [50]. However, the costs associated with constructing and maintaining these systems can be prohibitive for many small-scale farmers [37]. The initial capital outlay for RAS can exceed that of traditional aquaculture systems, making it economically vulnerable, especially during periods of unforeseen events such as power failures or disease outbreaks.

4.4.2 Economic viability

The long break-even periods associated with RAS investments further complicate their adoption [37]. Farmers may find it challenging to sustain operations during the extended timeframes required to recover initial costs. Also, RAS require significant energy to power the pumps, filters, and other components that drive the water recirculation and treatment processes [31]. This economic vulnerability is exacerbated by high operational costs related to energy consumption and technical management, which are critical for maintaining optimal conditions within RAS facilities. Consequently, many potential aquaculture entrepreneurs may be discouraged from adopting this technology due to perceived financial risks.

4.5 Technical barriers

4.5.1 Need for specialized skills

The successful implementation of RAS requires a skilled workforce capable of managing complex systems. Traditional aquaculture skills do not easily transfer to RAS operations, which necessitate specialized training and education. A significant barrier identified in various studies is the lack of knowledge among farmers regarding RAS technology, which hampers its widespread adoption [51]. For instance, research conducted among fish farmers in Haryana, India, revealed that 70% of respondents emphasized the critical need for specialized training in RAS operations [52]. This gap in expertise presents a significant hurdle for African aquaculture, where educational resources and training programs may be limited.

4.6 Operational challenges

In addition to the need for specialized skills, operational challenges associated with RAS systems can hinder their effectiveness. Issues such as system management, water quality control, and equipment maintenance require continuous monitoring and technical support [30]. The complexity of these systems means that even minor operational failures can lead to significant production losses [34]. For example, common technical challenges reported by RAS operators include difficulties with water temperature regulation and biofilter scaling. These operational hurdles necessitate ongoing investment in training and support services to ensure that farmers can effectively manage their systems.

5. Environmental challenges in traditional aquaculture

5.1 Pollution from waste

Traditional aquaculture systems often rely on open ponds or flow-through methods that discharge waste directly into the environment. This practice leads to significant pollution, primarily from uneaten feed, fish excrement, and chemicals used in farming operations [53]. As highlighted in various studies, the accumulation of organic waste in water bodies can result in eutrophication, a process that depletes oxygen levels and disrupts aquatic life [33]. For instance, nutrient overloads can trigger harmful algal blooms, which not only kill fish but also pose health risks to humans who consume contaminated seafood [54]. Moreover, the discharge of untreated wastewater can introduce pathogens and pollutants into local ecosystems, threatening biodiversity and the health of surrounding communities [37]. In many African countries, where water quality is already compromised by industrial and agricultural runoff, the additional burden from aquaculture waste exacerbates existing environmental problems [55]. The consequences of such pollution extend beyond immediate ecological impacts; they can also hinder the long-term viability of aquaculture by degrading the very resources upon which it depends.

5.2 Mitigating environmental impact with RAS

5.2.1 Advantages of RAS technology

Recirculating Aquaculture Systems (RAS) represent a transformative approach to aquaculture that addresses many of the environmental concerns associated with traditional methods. RAS operate on a closed-loop system that recycles water, significantly reducing freshwater consumption and minimizing waste discharge into the environment. According to research, RAS can recycle between 90% and 97% of the water used in fish farming operations [56]. This reduction in water usage is particularly beneficial in regions facing water scarcity. In addition to conserving water, RAS technology enhances environmental control by allowing for precise management of water quality parameters such as temperature, pH, and dissolved oxygen levels [57]. This level of control not only promotes optimal growth conditions for fish but also reduces stress on aquatic organisms, leading to healthier populations. Furthermore, RAS facilities are designed to filter out waste products effectively. Advanced filtration systems remove solids and convert toxic ammonia from fish waste into less harmful compounds through biological processes [58]. By capturing and managing waste within the system rather than discharging it into the environment, RAS significantly reduces the risk of pollution.

5.2.2 Economic and ecological benefits

The ecological advantages of RAS extend beyond pollution reduction; they also contribute to economic sustainability. By minimizing waste and optimizing resource use, RAS can lower operational costs for aquaculture producers. Additionally, because RAS facilities can be located near urban centers or markets, they reduce transportation emissions associated with seafood distribution [59]. This proximity not only decreases the carbon footprint but also enhances food security by providing fresh seafood year-round. Moreover, RAS technology supports biosecurity by isolating fish from wild populations [60]. This isolation minimizes disease transmission risks and reduces reliance on antibiotics or chemical treatments. Consequently, this approach aligns with growing consumer preferences for sustainably sourced seafood.

The environmental challenges posed by traditional aquaculture practices—particularly pollution from waste—are significant barriers to sustainable development in Africa's aquaculture sector. However, technologies like Recirculating Aquaculture

Systems offer viable solutions that mitigate these issues while enhancing productivity and profitability. By adopting RAS technology, African countries can promote more sustainable aquaculture practices that protect aquatic ecosystems and contribute to food security.

6. Case study in Africa - success stories

6.1 Wontesty ventures (Ghana)

Wontesty Ventures, founded in Ghana by Ing. Dr. Shadrack Kwadwo Amponsah, has emerged as a beacon of innovation and sustainability in the Ghanaian aquaculture industry. Wontesty Ventures is a pioneering business specializing in the setup of recirculating aquaculture fishponds, production, processing, value addition, and marketing of fish. With a vision to provide every household in Ghana with a pond by 2040, Wontesty Ventures aims to meet the growing local demand for fish while creating sustainable jobs for the youth and vulnerable communities. The company has introduced innovative fish pond technology, utilizing Recirculating Aquaculture Systems (RAS) instead of traditional dug-out ponds or cages. This technology allows for more efficient and sustainable fish farming practices [61]. Wontesty Ventures has grown from 100 tanks at inception to over 500 tanks nationwide, supporting 350 client farmers to produce 350 tons of catfish annually. The company has introduced what they call the Pond to Plate system (P2P). They have diversified their offerings beyond just raw catfish, now producing catfish fillets, nuggets, chunks, smoked catfish, and canned recipes. This initiative has renewed consumer interest and opened up new market opportunities for catfish products. Partnering with organizations such as the Bui Power Authority and GIZ, Wontesty Ventures has successfully implemented RAS setups across various regions, addressing challenges of water quality and availability. These interventions have not only enhanced production efficiency—achieving up to 30% increases in fish yield—but also supported livelihoods, created jobs, and promoted food security in local communities (**Figure 4**).

6.2 Victory farms (Kenya)

Victory Farms, a standout in Kenya's aquaculture sector, was founded in 2015 by Joseph Rehmann and Steve Moran. Over the years, it has become one of the leading tilapia producers in sub-Saharan Africa, driven by a mission to provide sustainable, high-nutrition protein to the African market [62]. The journey of Victory Farms is marked by impressive milestones: achieving an annual production of 10,000 tonnes of tilapia within 8 years, providing over 700 full-time jobs with wages significantly above the minimum wage, and establishing 75 retail outlets across Kenya to ensure efficient distribution of their fish. Furthermore, they have expanded their operations to Rwanda and set up a feed mill in Kenya. Their commitment to sustainability is noteworthy. Victory Farms aims to become carbon negative and actively engages with local communities. They provide high-speed rescue boats, fingerlings to local farmers, and tons of fish to children's homes. Additionally, their shareholder program for local landowners in Homa Bay ensures that the surrounding communities benefit from the farm's success (**Figure 5**).

Figure 4.
Overview of Wontesty Ventures Ghana, showcasing circular Recirculating Aquaculture Systems (RAS), farm layout, value-added fish products, and Intermediate Bulk Container (IBC) tanks for sustainable aquaculture.

Figure 5.
Overview of Victory Farms, Kenya, highlighting cage aquaculture systems on Lake Victoria, fish feeding operations, farm infrastructure, and team operations on boats, demonstrating sustainable fish farming practices.

6.3 Yalelo (Zambia)

Yalelo, a leading name in Zambia's aquaculture industry, was founded in 2011 by Phil Chaet and Allan Gray. Nestled in the pristine waters of Lake Kariba, Yalelo has swiftly risen to become one of the largest producers of freshwater fish in sub-Saharan Africa.

The company's vision to stabilize regional food security by developing sustainable fish protein resources has driven its impressive growth and innovation [63]. Yalelo has achieved remarkable milestones, producing 3000 tons of market-size tilapia annually despite the challenges of its remote location. This achievement is complemented by their significant contribution to the local economy, providing numerous job opportunities and supporting community development. With over 60 outlets across Zambia and neighboring countries, Yalelo ensures widespread accessibility to their high-quality fish. Sustainability is at the heart of Yalelo's operations. The company collaborates closely with local authorities and the government to address environmental sustainability, social responsibility, and traceability. Their sustainable aquaculture practices are designed to ease the pressure from decades of over-fishing, underscoring their commitment to environmental responsibility. Yalelo uses high-quality feed based on locally grown soy, maize, and wheat, ensuring that their fish are healthy and have a clean taste (**Figure 6**).

6.4 AgriMercarb limited (Ghana)

AgriMercarb Ghana, founded in 2022 by Luther Jesse Quarshie and partners, is revolutionizing the aquaculture landscape with its innovative approach to organic waste management. AgriMercarb has developed a network of community-based black soldier fly (BSF) production facilities across Ghana, significantly reducing organic waste while producing valuable resources [64]. This initiative includes transforming organic waste into protein-rich larvae for animal feed. The company's innovative approach also helps provide numerous job opportunities, fostering local economic development and community growth. The use of BSF larvae as fish feed is particularly noteworthy. BSF larvae are rich in protein and essential nutrients, making them an excellent alternative to traditional fish feed ingredients such as fishmeal and soy. This sustainable option reduces the dependency on overfished marine resources and provides a high-quality, cost-effective protein source for aquaculture.

Figure 6.
A collage of Yalelo Farms in Zambia, showcasing sustainable aquaculture practices, including fish feeding, net mending, cage farming, and local distribution.

By incorporating BSF larvae into fish feed, AgriMercarb helps fish farmers improve the growth rates and health of their fish, ultimately enhancing productivity and profitability in the aquaculture sector (**Figure 7**).

6.5 Chicoa fish farm (Mozambique)

Chicoa Fish Farm, established in 2011 and headquartered in Mozambique, is a pioneering force in Africa's aquaculture industry. Positioned on Lake Cahora Bassa, the company aims to tackle the region's protein deficit by producing high-quality, affordable farmed tilapia fish. Chicoa has achieved significant milestones, producing top-quality farmed tilapia with cost-efficient and environmentally sustainable methods that maintain a low carbon footprint [65]. The company aims to produce an additional million tonnes of fish to meet growing demand. By creating numerous job opportunities, Chicoa contributes significantly to local economic development and community growth. They supply fish to various markets, ensuring that their high-quality products are widely accessible. Sustainability is at the heart of Chicoa's operations. Their innovative offshore breeding methods reduce the need for onshore concrete ponds and make the production process independent of electricity. Partnering with the Department of Fisheries and development finance institutions, Chicoa trains and equips small-scale tilapia farmers across Mozambique, enhancing local farming practices and promoting sustainable development. The company's commitment to careful genetic selection and innovative breeding techniques highlights their dedication to excellence and sustainability (**Figure 8**).

6.6 Raanan fish feed (West Africa)

Raanan Fish Feed, founded in 1995 and headquartered in Israel, has emerged as a key player in Ghana's aquaculture industry. Established to revolutionize fish farming

Figure 7.
A collage of AgriMercarb Ltd. in Ghana, highlighting innovative insect-based animal feed production, featuring larvae cultivation and inspection of black soldier fly larvae.

Figure 8.
Chicoa Fish Farm's sustainable aquaculture operations, featuring cage systems, infrastructure, and collaborative efforts to enhance fish production and support local communities.

with superior fish feed, Raanan Fish Feed aims to support fish farmers with reliable, nutritious feed solutions that promote healthy growth and maximize yields. Raanan Fish Feed has made remarkable strides in the industry, notably by operating a state-of-the-art fish feed mill in Prampram, Ghana [66]. This facility, with an installed capacity of 24,000 metric tonnes per year, produces floating extruded fish feed for tilapia and catfish—the first of its kind in West Africa. This innovation has been instrumental in enhancing the quality and efficiency of fish farming in the region. The company supports numerous job opportunities, contributing to local economic development and community growth. They supply fish feed to over 500 farmers across Ghana and neighboring countries, ensuring widespread accessibility. The company collaborates closely with local authorities and international organizations to address environmental sustainability, social responsibility, and traceability. Their products are made from high-quality, locally sourced raw materials, and they offer technical assistance through workshops and training centers. This approach significantly impacts food security by providing an affordable supply of high-quality fish to the Ghanaian population (**Figure 9**).

6.7 AquaMet technologies (Ghana)

AquaMet Technologies, based in Ghana, is revolutionizing the aquaculture industry with its innovative technology solutions. Founded to address the challenges faced by fish farmers, AquaMet's mission is to provide affordable and sustainable tools that maximize yields and reduce uncertainties in aquaculture production. AquaMet has made significant strides in this field, notably through the development of a smart probe for monitoring water quality remotely *via* SMS, mobile app, and advisory services [67]. This technology helps reduce fish mortalities and increase yields by up to 35%. The company supports numerous job opportunities, fostering local economic development and community growth. Their AquaStall platform allows fish farmers to buy and sell fish farm inputs and high-quality fish from over 500 farmers, making

Figure 9.
Raanan Fish Feed's comprehensive operations, featuring high-quality fish feed products, bulk packaging, and efficient distribution systems supporting sustainable aquaculture and enhanced fish farming productivity.

their products widely accessible. Their smart probe and mobile app provide real-time advisory services, feed estimators, and record-keeping tools, helping farmers make informed decisions and improve their production processes. This approach significantly impacts food security by ensuring a steady supply of high-quality fish to the Ghanaian population (**Figure 10**).

6.8 AquaRech (Kenya)

AquaRech, founded in Kenya, is transforming the aquaculture industry with its innovative technology solutions aimed at empowering small and medium-scale fish farmers. Established to tackle the challenges fish farmers face in accessing quality inputs and markets, AquaRech's mission is to improve livelihoods and promote environmental sustainability through efficient fish farming methods. AquaRech has achieved significant milestones with the development of a mobile app that integrates IoT, farm management, and high-quality feeds to boost the productivity of small-scale fish farmers [68]. This app provides essential production metrics such as feed conversion ratio, average daily growth, mortality rates, and water temperature. By leveraging technology, AquaRech enables farmers to make informed decisions and optimize their farming practices, resulting in improved yields and profitability. The company creates numerous job opportunities, contributing to local economic development and community growth. AquaRech ensures the availability of high-quality fish feed and fingerlings through various outlets across Kenya, thereby enhancing food security and supporting regional development. These technological advancements, along with their high-quality feeds, set AquaRech apart in the industry, enabling farmers to remotely monitor water quality and improve production yields (**Figure 11**).

The success stories of these companies highlight the transformative potential of aquaculture in Africa. It demonstrates how innovation, sustainability, and community empowerment can drive growth and improve livelihoods in Africa.

Figure 10.
AquaMet's innovative solutions for aquaculture, including AquaStall Marketplace for fish and farm inputs, the AquaMet Smart Probe for water quality monitoring, and the AquaMet Mobile App for farm management and advisory services.

Figure 11.
Overview of AquaRech's innovative aquaculture solutions in Kenya, showcasing smart farming technology, farmer engagement, and operational hubs.

7. Investment and policy opportunities

Investment and policy opportunities in the aquaculture sector are critical for enhancing food security, promoting sustainable practices, and fostering economic growth in Africa. Governments and private sectors can collaborate effectively to increase investment in aquaculture infrastructure and capacity-building programs.

DOI: http://dx.doi.org/10.5772/intechopen.1008743

8. Strategic partnerships

One of the most effective ways to increase investment in aquaculture is through strategic partnerships between governments, private investors, and non-governmental organizations (NGOs). These collaborations can facilitate knowledge transfer, technology sharing, and resource mobilization. For instance, government bodies can partner with private sector players to develop aquaculture parks that provide the necessary infrastructure for fish farming, such as hatcheries, processing facilities, and distribution networks. Such initiatives not only enhance production capabilities but also create jobs and stimulate local economies. Additionally, international partnerships can play a significant role in bringing expertise and funding to local aquaculture projects. Programs like the AquaFish Innovation Lab illustrate how collaborative efforts can build institutional capacity and improve aquaculture practices through training and outreach activities. By engaging stakeholders from both public and private sectors in these initiatives, countries can leverage global best practices to enhance their local aquaculture industries.

9. Financial support mechanisms

Governments can also implement financial support mechanisms to encourage private investment in aquaculture. This could include providing low-interest loans, grants, or tax incentives for businesses that invest in sustainable aquaculture practices. For example, Ghana's recent policy initiatives aim to boost aquaculture production significantly by increasing the market share of commercially farmed fish from 14% in 2021 to 25% by 2027. Such policies are designed to attract both local and foreign investments into the sector. Furthermore, establishing public-private investment funds dedicated to aquaculture development can help mitigate financial risks for investors. These funds could be used to support research and development initiatives that focus on innovative technologies such as recirculating aquaculture systems (RAS), which require less water and land compared to traditional methods. By demonstrating the economic viability of sustainable practices through pilot projects funded by these investments, governments can encourage broader participation from the private sector.

10. Capacity-building initiatives

Capacity-building programs are essential for ensuring that stakeholders have the necessary skills and knowledge to implement effective aquaculture practices. Governments should prioritize training programs that equip farmers with the best management practices in fish farming, water quality monitoring, and disease management. Collaborations with institutions like the Food and Agriculture Organization (FAO) can provide valuable resources and training materials tailored to local needs. Moreover, engaging local communities in capacity-building efforts fosters ownership of aquaculture projects and promotes sustainable practices. Training programs should be designed to be inclusive, providing equal opportunities for men and women to participate in aquaculture activities. This approach not only enhances community resilience but also contributes to gender equality within the sector.

The collaboration between governments and the private sector is vital for increasing investment in aquaculture infrastructure and capacity-building programs in Africa. By forming strategic partnerships, implementing financial support mechanisms, and prioritizing capacity-building initiatives, stakeholders can create a robust aquaculture industry that meets the growing demand for fish while promoting sustainability. As countries like Ghana demonstrate through their policy frameworks, a concerted effort toward enhancing aquaculture can lead to significant economic benefits and improved food security across the continent.

11. Conclusion

Aquaculture in Africa holds immense potential to address critical challenges of food security, economic development, and environmental sustainability. Despite numerous hurdles, including resource constraints, financial barriers, and limited technical expertise, the sector has demonstrated promising growth fueled by innovative practices and strategic partnerships. Traditional systems, integrated approaches, and advanced technologies such as recirculating aquaculture systems (RAS) are transforming the industry, providing sustainable solutions to optimize resource efficiency, enhance productivity, and reduce environmental impact.

Success stories across the continent underscore the transformative power of aquaculture, with enterprises such as Wontesty Ventures, Victory Farms, and Yalelo leading the charge in innovation, community empowerment, and sustainability. These examples highlight the role of collaboration, capacity-building, and policy support in unlocking the potential of aquaculture to benefit both rural and urban populations.

As Africa's population continues to grow, the aquaculture industry must adapt to evolving demands by leveraging technology, fostering public-private partnerships, and scaling sustainable practices. By addressing key challenges and embracing opportunities, aquaculture can play a pivotal role in achieving food and nutrition security while promoting economic resilience and ecological balance for future generations.

Acknowledgements

The authors acknowledge the use of Notebook LM, Perplexity AI, and ChatGPT for language polishing of the manuscript.

Aquaculture in Africa: Production Systems, Challenges, and Opportunities
DOI: http://dx.doi.org/10.5772/intechopen.1008743

Author details

Shadrack Kwadwo Amponsah[1*], Divine Dogbeda Azumah[1]
and Seth Koranteng Agyakwah[2]

1 CSIR – Crops Research Institute, Department of Agricultural Engineering and Transport, Ghana, Africa

2 CSIR – Water Research Institute, Department of Fisheries and Aquaculture, Ghana, Africa

*Address all correspondence to: skamponsah@hotmail.com

References

[1] Henchion M, Hayes M, Mullen A, Fenelon M, Brijesh KT. Future protein supply and demand: Strategies and factors influencing a sustainable equilibrium. Food. 2017;**6**:53

[2] Asiedu B, Nunoo FKE, Iddrisu S. Prospects and sustainability of aquaculture development in Ghana, West Africa. Cogent Food & Agriculture [Internet]. 2017;**3**(1):1-13. DOI: 10.1080/23311932.2017.1349531

[3] Cowx IG, Ogutu-Owhayo R. Towards sustainable fisheries and aquaculture management in the African Great Lakes. Fisheries Management and Ecology. 2019;**26**(5):397-405

[4] Obiero K, Meulenbroek P, Drexler S, Dagne A, Akoll P, Odong R, et al. The contribution of fish to food and nutrition security in eastern Africa: Emerging trends and future outlooks. Sustainability. 2019;**11**(6):1-15

[5] FAO. The State of World Fisheries and Aquaculture. Vol. 32. Rome. Italy: FAO; 2020. 244 p

[6] Hinrichisen E, Walakira JK, Langi S, Ibrahim NA, Tarus V, Badmus O, et al. Prospects for Aquaculture Development in Africa: A Review of Past Performance to Assess Future Potential. ZEF Working Paper Series, No. 211. Bonn: University of Bonn, Center for Development Research (ZEF); 2022. p. 50

[7] FAO. FAO Report: Global Fisheries and Aquaculture Production Reaches a New Record High, Untapped Potential Remains in Africa. Rome, Italy: FAO [Internet]; 2024. Available from: https://www.fao.org/africa/news-stories/news-detail/fao-report--global-fisheries-and-aquaculture-production-reaches-a-new-record-high--untapped-potential-remains-in-africa/en

[8] Adeleke B, Robertson-Andersson D, Moodley G, Taylor S. Aquaculture in Africa: A comparative review of Egypt, Nigeria, and Uganda Vis-À-Vis South Africa. Reviews in Fisheries Science & Aquaculture. 2020;**29**(2):167-197. DOI: 10.1080/23308249.2020.1795615

[9] Ottinger M, Clauss K, Kuenzer C. Opportunities and challenges for the estimation of aquaculture production based on earth observation data. Remote Sensing. 2018;**10**(7):1-24

[10] Manu EK, Chen GS, Hoang N, Leu S. Natural resource extraction and environmental sustainability in Africa: The role of voice and accountability. Sustainable Development. 2024;**32**:1-16

[11] Kaleem O, Sabi BS, AF. Overview of aquaculture systems in Egypt and Nigeria, prospects, potentials, and constraints. Aquaculture and Fisheries [Internet]. 2021;**6**(6):535-547. DOI: 10.1016/j.aaf.2020.07.017

[12] Brummett RE, Lazard J, Moehl J. African aquaculture: Realizing the potential. Food Policy. 2008;**33**(5):371-385

[13] Duijn AP, van der Heijden GMP, Bolman B, Rurangwa E. Review and analysis of small-scale aquaculture production in East Africa. Wageningen Centre for Development Innovation [Internet]. 2018;**48**:1-34. Available from: www.wur.eu/cdi

[14] Rothuis A, Duijn AP, Van Roem A, Ouwehand A, Van Der PW, Rurangwa E. Aquaculture business opportunities in Egypt. LEI Report, IMARES Report [Internet]. 2013;**39**:2013-2039. Available from: http://edepot.wur.nl/258663

[15] Okwodu NE. Aquaculture for sustainable development in

Nigeria. World Scientific News. 2016;**47**(2):1-13. Available from: www.worldscientificnews.com

[16] Amenyogbe E, Chen G, Wang Z, Lin M, Lu X, Atujona D, et al. A review of Ghanas aquaculture industry. Journal of Aquaculture Research and Development. 2018;**9**(8):1-6

[17] Olaoye OJ, Adegbite DA, Oluwalana EO, Vaughan IO, Odebiyi CO, Adediji AP. Comparative evaluation of economic benefits of earthen fish ponds and concrete tanks in aquaculture enterprises in Oyo state, Nigeria. Ribarstvo, Croatian Journal of Fisheries. 2014;**72**(3):107-117

[18] Anetekhai MA, Akin-Oriola GA, Aderinola OJ, Akintola SL. Steps ahead for aquaculture development in sub-Saharan Africa - the case of Nigeria. Aquaculture. 2004;**239**(1-4):237-248

[19] Reginald Inodu K, James E. Comparative assessment of growth performance and economics of production of Clarias gariepinus fingerlings in ponds and tanks. Greener Journal of Agricultural Sciences. 2014;**4**(2):034-038

[20] Akuffo J. University of Cape Coast. Ghana, 2017;(November):9-10

[21] Clough S, Mamo J, Hoevenaars K, Bardocz T, Petersen P, Rosendorf P, et al. Innovative technologies to promote sustainable recirculating aquaculture in eastern Africa—A case study of a Nile tilapia (*Oreochromis niloticus*) hatchery in Kisumu, Kenya. Integrated Environmental Assessment and Management. 2020;**16**(6):934-941

[22] Jamu DM, Ayinla OA. Potential for the development of aquaculture in Africa. Naga, The WorldFish Center Quarterly. 2003;**26**(3):9-13

[23] Agyei BP. Sustainability of Pond and Cage Fish Farming Systems in the Ashanti, Bono, Eastern and Volta Regions in Ghana [Thesis]. Ås, Norway: Norwegian University of Life Sciences, Norway; 2022. Available from: https://www.nmbu.no/fakultet/landsam/institutt/noragric

[24] Odei DK. Sustainable Development of Aquaculture on the Volta Lake- a Case Study of the Asuogyaman District in the Eastern Region of Ghana. Vol. May. Norway: UiT The Arctic University of Norway; 2015

[25] Magna EK, Mensah ETD, Mabe FN, Johnson-Ashun M, Konadu LO, Appiah EK. Profitability analysis of small-scale cage aquaculture farms in the Volta Lake of Ghana. Aquaculture Research. 2023;**2023**:1-10

[26] Kassam L, Dorward A. A comparative assessment of the poverty impacts of pond and cage aquaculture in Ghana. Aquaculture. 2017;**470**:110-122

[27] Prein M. Integration of aquaculture into crop - animal systems in Asia. Agricultural Systems. 2002;**71**(1-2):127-146

[28] Shaalan M, El-Mahdy M, Saleh M, El-Matbouli M. Aquaculture in Egypt: Insights on the current trends and future perspectives for sustainable development. Reviews in Fisheries Science & Aquaculture. 2018;**26**(1):99-110. DOI: 10.1080/23308249.2017.1358696

[29] Badiola M, Mendiola D. Keys to Energy Efficiency in RAS. Ontario, Canada: Hatchery International [Internet]; 2018. Available from: https://www.hatcheryinternational.com/power-struggle-keys-to-energy-efficiency-in-ras-3299

[30] Lindholm-Lehto P. Water quality monitoring in recirculating aquaculture

systems. Aquaculture, Fish and Fisheries. 2023;**3**(2):113-131

[31] Murray F, Bostock J, Fletcher D. Review of recirculation aquaculture system technologies and their commercial application prepared for Highlands and Islands Enterprise. 2014;**44**(March):1-82. Available from: http://hdl.handle.net/1893/21109

[32] Newase R. Exploring Recirculating Aquaculture Systems: Efficient Use of Water and Resources. Mumbai, India: Organica Biotech [Internet]; 2024. Available from: https://organicabiotech.com/exploring-recirculating-aquaculture-systems-efficient-use-of-water-and-resources/

[33] Zhang SY, Li G, Wu HB, Liu XG, Yao YH, Tao L, et al. An integrated recirculating aquaculture system (RAS) for land-based fish farming: The effects on water quality and fish production. Aquacultural Engineering. 2011;**45**(3):93-102

[34] Benjamin EO, Ola O, Buchenrieder GR. Feasibility study of a small-scale recirculating aquaculture system for sustainable (Peri-)urban farming in sub-Saharan Africa: A Nigerian perspective. Land. 2022;**11**(11):1-16

[35] Timmons M, Guerdat T, Vinci B. Recirculating Aquaculture. New York: Ithaca Publishing Company LLC; 2018

[36] Segura J. Recirculating aquaculture systems (RAS): A solution for water efficiency. Journal of Biodiversity, Bioprospecting and Development. 2023;**9**:2

[37] Verdegem M, Buschmann AH, Latt UW, Dalsgaard AJT, Lovatelli A. The contribution of aquaculture systems to global aquaculture production. Journal of the World Aquaculture Society. 2023;**54**(2):206-250

[38] Becke C, Schumann M, Steinhagen D, Rojas-Tirado P, Geist J, Brinker A. Effects of unionized ammonia and suspended solids on rainbow trout (*Oncorhynchus mykiss*) in recirculating aquaculture systems. Aquaculture [Internet]. 2019;**499**:348-357. DOI: 10.1016/j.aquaculture.2018.09.048

[39] Aich N, Nama S, Biswal A, Paul T. A review on recirculating aquaculture systems: Challenges and opportunities for sustainable aquaculture. Inno Farms [Internet]. 2020;**5**(1):17-24. Available from: www.innovativefarming.in

[40] Strauch SM, Wenzel LC, Bischoff A, Dellwig O, Klein J, Schüch A, et al. Commercial African catfish (*Clarias gariepinus*) recirculating aquaculture systems: Assessment of element and energy pathways with special focus on the phosphorus cycle. Sustainability. 2018;**10**(6):1-29

[41] Munguti JM, Kim JD, Ogello EO. An overview of Kenyan aquaculture: Current status, challenges, and opportunities for future development. Fisheries and Aquatic Science. 2014;**17**(1):1-11

[42] Suloma A, Ogata HY. Future of rice-fish culture, desert aquaculture and feed development in Africa: The case of Egypt as the leading country in Africa. The Japan Agricultural Research Quarterly. 2006;**40**(4):351-360

[43] Walakira JK, Hinrichsen E, Tarus V, Langi S, Ibrahim NA. Scaling aquaculture for food security and employment in Africa: Insights from Egypt, Kenya and Nigeria. Working Paper Series. 2023;**June**:1-38

[44] Halwart M, van Dam AA. Integrated Irrigation and Aquaculture in West Africa Concepts, Practices and Potentials. Rome. Italy: FAO; 2006. 197 p

[45] Siqwepu O, Salie K, Goosen N. Evaluation of potassium diformate and potassium chloride in the diet of the African catfish, *Clarias gariepinus* in a recirculating aquaculture system. Aquaculture. 2020;**526**:1-41

[46] Gabriel UU, Akinrotimi O, Bekibele DO, Onunkwo DN, Anyanwu PE, Harcourt P. Locally produced fish feed: Potentials for aquaculture development in sub-Saharan Africa. African. Journal of Agricultural Research. 2007;**2**(7):287-295

[47] Mahieu A. Fish-Farming in South Africa: A Study of the Market Environment and the Suitable Species. Vol. March. Stellenbosch: Stellenbosch University; 2015. p. 77

[48] Edwards P. Aquaculture environment interactions: Past, present and likely future trends. Aquaculture [Internet]. 2015;**447**:2-14. DOI: 10.1016/j.aquaculture.2015.02.001

[49] Nazar AKA, Jayakumar R, Tamilmani G. Recirculating aquaculture systems. Tamil Nadu, India: Winter School on Technological Advances in Mariculture for Production Enhancement and Sustainability; 2012. pp. 136-139

[50] Amponsah SK, Guilherme L. Promoting Domestic Production of Fish Using Recirculating Aquaculture System (RAS). London, UK: IntechOpen [Internet]; 2021. p. 13. DOI: 10.1016/j.colsurfa.2011.12.014

[51] Paruğ Ş, Jabrr A, Lazrag FAH. Sustainable development of fisheries and aquaculture in North Africa: Current status and future prospects. Menba Kastamonu University Faculty of Fisheries Journal. 2024;**10**(1):90-105

[52] Singh P, Gulati R, Sharma R, Lowansanhi A. Constraints and challenges of fish farmers with a recirculatory aquaculture system (RAS) in Haryana state, India. Environment Conservation Journal. 2024;**25**(2):483-487

[53] Ragasa C, Andam KS, Kufoalor DS, Amewu S. A blue revolution in sub-Saharan Africa? Evidence from Ghana's tilapia value chain. 2018;**June**:31

[54] Lal J, Vaishnav A, Verma DK, Jana A, Jayaswal R, Chakraborty A, et al. Emerging innovations in aquaculture: Navigating towards sustainable solutions. International Journal of Environment and Climate Change. 2024;**14**(7):83-96

[55] Jagger P, Pender J. Markets, marketing and production issues for aquaculture in East Africa: The case of Uganda. ICLARM Quarterly [Internet]. 2001;**24**(June):1-10. Available from: http://aquaticcommons.org/9244/

[56] Grieg Sea Food. Recirculating Aquaculture Systems (RAS) [Internet]. Bergen, Norway: Grieg Sea Food; 2020. Available from: https://griegseafood.com/our-impact-recirculating-aquaculture-systems

[57] Mugwanya M, Dawood MAO, Kimera F, Sewilam H. A review on recirculating aquaculture system: Influence of stocking density on fish and crustacean behavior, growth performance, and immunity. Annals of Animal Science. 2022;**22**(3):873-884

[58] Palm HW, Knaus U, Wasenitz B, Bischoff AA, Strauch SM. Proportional up scaling of African catfish (*Clarias gariepinus* Burchell, 1822) commercial recirculating aquaculture systems disproportionally affects nutrient dynamics. Aquaculture. 2018;**491**:155-168

[59] Badiola M. The RAS Case for Sustainability. Ontario, Canada:

Hatchery International [Internet]; 2019. Available from: https://www.hatcheryinternational.com/power-struggle-sense-of-sustainability-3446/

[60] Derwent Group. Main Advantages of Recirculating Aquaculture Systems. Cantabria, Spain: Derwent Group; 2019. Available from: https://derwent.es/en/advantages-recirculating-aquaculture-systems/

[61] VC4A. Wontesty Ventures [Internet]. 2021. Available from: https://vc4a.com/ventures/wontesty-ventures/

[62] Farms V. Company - Victory Farms [Internet]. 2023. Available from: https://victoryfarmskenya.com/

[63] Group F. Yalelo Zambia's #1 Fish! [Internet]. 2024. Available from: https://yalelo.com/

[64] Agrimercarb. Company - Agrimercarb Ltd [Internet]. 2023. Available from: https://agrimercarb.com/company/

[65] Farms CF. Rethink the potential of African Aquaculture: Chicoa Fish Farms [Internet]. 2025. Available from: https://chicoa.fish/

[66] Feed RF. Ranaan Fish feed West Africa [Internet]. 2022. Available from: https://www.raananfishfeed-wa.com/

[67] Limited AT. AquaMet Technologies [Internet]. p. 2023. Available from: https://www.aquamettechnologies.com/

[68] Aquarech. Africa's first fish farming platfrom: Transitioning small and medium sized fish farmers to profitability [Internet]. 2023. Available from: https://www.aquarech.com/

Chapter 3

New Perspectives in Fisheries: The Use of Insects in Aquaculture

Isaac Osuga, Catherine Maindi, Vincent Mwashi and Chrysantus Tanga

Abstract

In the face of climate change and variability and the need to enhance aquaculture production sustainability, production and utilization of novel feed resources for aquaculture while maintaining or contributing to environmental sustainability is critical. Insects have been shown to produce critical biomass suitable for animal feed with minimal environmental footprints. The insect biomass has been shown to be of high nutritional quality and therefore can be used as feed for fish. Fish feed formulations have been successfully done and incorporated diets for various fish species with very positive results. The incorporation of the insect meals in aquafeeds has also been shown to reduce the cost of fish feeds and improve the overall profitability of fish farming enterprises. In this chapter, the utilization of insect meals in the formulation of aquafeeds and the effect on the performance of fish is presented. This includes the replacement of fishmeal as the main animal protein source in fish feeds and the nutritional quality of insect meals as important sources of proteins for green, profitable, and sustainable aquaculture. It is certain that in the near future, large-scale insect farming and processing to produce insect meals as an ingredient of fish feeds will have positive impact on the sustainability and profitability of aquaculture.

Keywords: insects for aquafeed, insect meals, climate change, fish feeding, feed formulation

1. Introduction

The global demand for fish protein has been steadily increasing, driven by factors such as population growth, rising consumer preference for fish and its associated health benefits [1], urbanization, and general economic growth. This significant increase in demand has contributed to the rapid development of world fisheries, particularly the intensification and capitalization of aquaculture. The Food and Agriculture Organization of the United Nations (FAO) estimates that aquaculture will play an increasingly important role in meeting this demand, contributing a significant portion of the world's seafood supply [2]. As wild fish stocks dwindle, aquaculture seems to be the only option to ensure a sustainable supply of high quality (animal) protein for a growing population [3]. For instance, the global food fish consumption increased at an average rate of 3.0% per year from 1961 to 2021, which is almost twice the rate of annual world population growth, which is 1.6% for the same period [2]. Additionally, aquaculture emerges as a vital protein source not only for humans, but also for animal consumption [4]. Aquaculture represents a crucial means of meeting the increasing demand for food while addressing nutritional deficiencies and having a lower environmental impact than many other animal-based food sources except insect food and feed sources.

The provision of sustainable aquafeed is a crucial component of successful aquaculture production, as it is one of the most significant inputs. Ensuring a reliable, sustainable, and low-cost feed supply is essential for supporting the future growth of global aquaculture production. Developing effective, sustainable feed formulations is key to achieving this goal. Aquaculture is the primary consumer of fishmeal and fish oil, which are derived from small, oceanic fish. For instance, in 2017, 18.3% of global marine fish catch was utilized as feed ingredients for aquaculture [5]. In 2018, global aquaculture feed production reached 40.1 million metric tons, demonstrating a 4% increase from the previous year [6]. Aquaculture feeds play a crucial role in supporting fish growth, survival, and overall health. The quality and composition of aquaculture feeds significantly influence fish performance [7].

Traditionally, fishmeal, derived from wild-caught fish, has been the primary protein source in commercial fish feeds in aquaculture. Fishmeal is a nutrient-dense feed ingredient used in aquaculture, providing high-quality protein with a balanced amino acid profile, essential omega-3 fatty acids, vitamins, minerals, and trace elements. Its high digestibility and palatability make it an ideal feed for aquatic animals [8]. However, the traditional methods of fishmeal production, relying on wild-caught fish, have placed significant pressure on wild fish stocks, thus facing sustainability challenges. Overfishing has depleted fish stocks in many regions [3]. The escalating costs of aquaculture production, primarily driven by the rising price of fishmeal and fish oil, jeopardize the sector's sustainability [9]. This overreliance on marine-derived feed ingredients, coupled with declining production and increased competition for limited resources, has led to significant price volatility and reduced availability of conventional feedstuffs [10]. Moreover, most of these fish are suitable for human consumption [11]. Furthermore, as climate change and variability intensify, there is an urgent need to innovate food/feed production systems. By adopting new approaches and phasing out less sustainable practices, we can conserve resources, reduce greenhouse gas emissions, and enhance carbon sequestration [12, 13].

Given the scarcity of fishmeal, researchers have explored alternative protein sources that offer similar nutritional benefits [14]. Consequently, there has been a concerted effort to find alternative protein sources that can support the aquaculture industry's economic and environmental goals [15]. This has necessitated the development of alternative, sustainable, and cost-effective feed ingredients to maintain the quality and nutritional value of farmed fish. A variety of alternative feed sources, including plant-based and insect-based proteins, are being investigated. Plant-based materials like soybeans, oilseeds, and cereal gluten are increasingly incorporated into animal feeds [14]. However, replacing fishmeal with large amounts of plant-based ingredients and fish oil is not practical in aquaculture. Plant-based feeds often contain anti-nutritional factors, non-starch polysaccharides, and less optimal fatty acid and amino acid profiles for fish [14]. Plant-based ingredients have been associated with negative gut health effects in carnivorous fish due to the presence of anti-nutritional factors and complex carbohydrates [16]. According to various researchers [17–19], incorporating high levels of soybean meal in various carnivorous fish diets has been associated with negative impacts on growth, gut health, liver function, gut microbiota, and immune response. This can negatively impact fish growth and welfare. As a result, researchers have sought to develop innovative aquafeeds that can replace fishmeal while minimizing the adverse effects of plant-based proteins [4]. Recent research has focused on exploring sustainable alternative feed sources such as insects [20, 21]. Insects have emerged as a promising and widely applicable option for aquaculture.

Insect-based diets have gained significant attention as a sustainable and efficient alternative to fishmeal. Given the limitations of fishmeal, insects offer a promising alternative due to their similar nutritional profile and potential for sustainable production [4]. Insects are a rich source of protein, with crude protein content ranging from 34 to 74% dry matter [22]. Whole insects typically contain 42 to 63.3% crude protein (DM), which can increase to up to 74% DM when defatted [4]. Additionally, insects offer a balanced amino acid profile similar to fishmeal, high levels of lipids and essential fatty acids, vitamins (e.g., vitamin B_{12}), and minerals (e.g., iron and zinc) [22]. Furthermore, insect meals contain bioactive compounds such as chitin, fatty acids, anti-oxidants, and antimicrobial peptides, which can promote animal health and combat antimicrobial resistance [23, 24]. The nutritional composition of insects can vary based on species, rearing substrates and conditions, and processing methods [4, 25]. For example, defatting insects can increase their protein content, while feeding them diets rich in omega-3 fatty acids can enhance their fatty acid profile [25]. Unlike fishmeal and plant-based proteins, insects can be produced intensively, efficiently, and sustainably, with minimal environmental impact [22]. When combined with plant-based proteins, insects can mitigate the negative effects on fish growth and gut health often associated with high plant-protein diets [16, 26]. Insects are characterized by rapid growth, efficient feed conversion, and the ability to be reared on organic waste products [27].

A variety of insect species have been explored for use in aquaculture feeds. Among the numerous insects that have emerged as the most promising species explored for aquaculture feeds include: silkworms (*Bombyx mori*), black soldier flies (*Hermetia illucens*), houseflies (*Musca domestica*), yellow mealworms (*Tenebrio molitor*), lesser mealworms (*Alphitobius diaperinus*), house crickets (*Acheta domesticus*), banded crickets (*Gryllodes sigillatus*), and Jamaican field crickets (*Gryllus assimilis*) [15, 28, 29]. These species have been extensively studied as potential replacements for fishmeal in aquaculture feeds and have been approved for use in the European Union [14] and many parts of the world. For the past two decades, researchers have investigated the potential of insect meals as a sustainable alternative to fishmeal in aquaculture [14]. Numerous studies have explored insect species identification, cultivation methods, nutritional value, large-scale production, and safety concerns [30]. Recent feeding trials have demonstrated the feasibility of partially replacing fishmeal with insect meals in various aquaculture species, although excessive substitution can negatively impact growth [31, 32]. The growing interest in insect-based aquaculture feeds has led to increased investment in insect production and the development of innovative feed formulations. This chapter therefore examines the most promising insect species for innovative aquaculture feed production, considering their nutritional value, ease of rearing, biomass production, and economic feasibility potential.

2. Nutritional composition of insects utilized in aquafeeds

Insects, often overlooked as a potential food source, have emerged as a promising alternative for aquaculture feeds. Their high protein content is well-balanced in essential amino acids, including lysine, methionine, threonine, and essential fatty acids, which are crucial for fish growth and development [22]. The amino acid profiles of insects are often superior to those of plant-based protein sources like soybean meal. By incorporating insect-based ingredients into fish feeds, the aquaculture industry can reduce its environmental impact, enhance the nutritional value of farmed fish, and improve the overall sustainability of aquaculture practices. Meta-analysis, a powerful statistical technique, has been widely employed to synthesize existing research on fish nutrition, including amino acid and mineral requirements [33], and the impact of replacing fishmeal with alternative protein sources [34]. Numerous studies have explored the nutritional value of various insect species and their potential use in aquaculture feed [35]. The nutritional composition of insects has been well-documented in numerous studies [15, 27, 36, 37]. However, it is crucial to determine the specific nutritional profile of an insect species, as it can vary based on factors such as life stage, rearing conditions, and rearing substrate. This information is essential for ensuring that the insect meal meets the specific dietary requirements of the target fish species.

To date, at least 16 insect species have been evaluated as potential protein sources for aquaculture feed [38, 39]. Among these numerous insects, eight of them have emerged as the most promising species explored for aquaculture feeds and include silkworms (*Bombyx mori*), black soldier flies (*Hermetia illucens*), houseflies (*Musca domestica*), yellow mealworms (*Tenebrio molitor*), lesser mealworms (*Alphitobius diaperinus*), house crickets (*Acheta domesticus*), banded crickets (*Gryllodes sigillatus*), and Jamaican field crickets (*Gryllus assimilis*) [15, 28, 29]. Extensive research has been conducted to characterize the nutritional composition of these insects, including protein content, amino acid profiles, fat content, fatty acid profiles, vitamin, and mineral content [27, 37, 38]. **Table 1** presents the primary chemical constituents of various insect species.

2.1 Protein and amino acids

Fish species have varying protein requirements, ranging from 28 to 55% of dry diet, with the highest demand occurring during the larval and fry stages, particularly

	Crude Protein	Lipids	Calcium	Phosphorus	Ca:P ratio
Black soldier fly larvae	42.1 (56.9)*	26.0	7.6	0.9	8.4
Housefly maggot meal	50.4 (62.1)	18.9	0.5	1.6	0.3
Mealworm meal	52.8–82.6	36.1	0.3	0.8	0.4
Locust meal	57.3 (62.6)	8.5	0.1	0.1	1.2
House cricket	63.3 (76.5)	17.3	1.0	0.8	1.3
Mormon cricket	59.8 (69.0)	13.3	0.2	1.0	0.2
Silkworm pupae meal	60.7 (81.7)	25.7	0.4	0.6	0.6
Fishmeal	70.6	9.9	4.3	2.8	1.6
Soybean meal	51.8	2	0.4	0.7	0.6

**Values in parentheses are calculated values of the defatted meals.*

Table 1.
Main chemical constituents (% dry matter) in insect meals, fishmeal, and soybean meal.

for carnivorous fish [40]. Plant-based feedstuffs often lack sufficient protein and essential amino acids, particularly methionine and lysine. Deficiencies in these amino acids can lead to increased feed intake, reduced growth, poor feed conversion efficiency, and increased susceptibility to diseases [41]. Insects are a rich protein source, with protein content ranging from 25 to 75% of their dry weight [42, 43]. Proteins are composed of amino acids, and the total protein content is equivalent to the sum of these amino acids. Compared to the larvae of black soldier flies, houseflies, and mealworm, adult pupae of silkworms, locusts, and crickets exhibit higher levels of crude protein [44]. All fish species require essential amino acids (EAAs) for optimal growth [40]. However, most insect species studied for fish diets, including silkworms (*Bombyx mori*), yellow mealworms (*Tenebrio molitor*), and house crickets (*Acheta domesticus*), exhibit amino acid profiles that align well with fish requirements (**Table 2**) [15, 37, 40]. Silkworms, in particular, are notable for their high methionine content [37, 45]. Silkworms, black soldier flies, and houseflies exhibit a superior amino acid profile compared to soybean meal. Consequently, these insects are more suitable replacements for fishmeal in aquaculture feeds [36, 38]. Black soldier flies, houseflies, and silkworms are relatively rich in lysine [4]. In comparison with fishmeal, locusts, crickets, and mealworms exhibit lower lysine levels. While threonine levels are similar across most insect species, silkworms have a higher concentration [36, 38]. Except for silkworms and housefly maggots, the other six insect species have lower tryptophan levels compared to fishmeal. To ensure optimal growth, supplemental synthetic amino acids may be necessary, depending on the specific dietary needs of the fish species [4].

2.2 The lipid content and fatty acid

Fatty acids, the building blocks of fats, are categorized as saturated, monounsaturated, or poly-unsaturated based on their degree of saturation [46]. Polyunsaturated fatty acids are further classified into omega- 3, 6, or 9 fatty acids. Cold-water carnivorous fish, like salmon, can tolerate diets with up to 35% lipid and a protein-to-lipid ratio as low as 16:1 [47]. In contrast, warm-water fish require a higher protein-to-lipid ratio (25–26:1) [47], and herbivorous and omnivorous warm-water fish have lower tolerance for dietary lipids. High dietary lipid levels (20%) can compromise the immune function of some marine omnivorous species [48]. Compared to fishmeal (8.2%) and soybean meal (3.0%), insects generally have higher lipid content, ranging from 10 to 30%. Most of the insect species used in aquaculture production exhibit lower fat content compared to fishmeal. Insects accumulate fat, particularly during their embryonic stages [36]. The fat content varies among species, ranging from approximately 8% in mature locusts to 36% in mealworm larvae [49]. It is important to note that lipid content can fluctuate within a species due to factors such as developmental stage and diet [50]. Males typically have lower fat reserves than females [51]. Fish oil contains significantly higher levels of omega-3 fatty acids compared to insect meals [27]. In contrast, insect meals are rich in saturated fatty acids. The specific lipid and fatty acid profile of insects is influenced by their diet [15]. While freshwater fish require poly-unsaturated fatty acids (PUFAs), marine fish need highly unsaturated fatty acids (HUFAs) in their diet [52]. Although house crickets, mealworms, and housefly maggots exhibit higher levels of unsaturated fatty acids (around 60–70%), black soldier fly larvae have relatively low levels of unsaturated fatty acids (approximately 19–37%) [22, 49]. Insects that have been utilized in fish feeding are rich in poly-unsaturated fatty acids (PUFAs),

Amino acids	Black soldier fly larvae	Housefly maggot meal	Mealworm	Locust meal	House cricket	Mormon cricket	Silkworm pupae meal	Fishmeal	Soybean meal
Essential									
Methionine	2.1	2.2	1.5	2.3	1.4	1.4	3.5	2.7	1.3
Cystine	0.1	0.7	0.8	1.1	0.8	0.1	1.0	0.8	1.4
Valine	8.2	4.0	6.0	4.0	5.1	6.0	5.5	4.9	4.5
Isoleucine	5.1	3.2	4.6	4	4.4	4.8	5.1	4.2	4.2
Leucine	7.9	5.4	8.6	5.8	9.8	8.0	7.5	7.2	7.6
Phenylalanine	5.2	4.6	4.0	3.4	3.0	2.5	5.2	3.9	5.2
Tyrosine	6.9	4.7	7.4	3.3	5.2	5.2	5.9	3.1	3.4
Histidine	3.0	2.4	3.4	3.0	2.3	3.0	2.6	2.4	3.1
Lysine	6.6	6.1	5.4	4.7	5.4	5.9	7.0	7.5	6.2
Threonine	3.7	3.5	4.0	3.5	3.6	4.2	5.1	4.1	3.8
Tryptophan	0.5	1.5	0.6	0.8	0.6	0.6	0.9	1.0	1.4
Non-essential									
Serine	3.1	3.6	7.0	5	4.6	4.9	5.0	3.9	5.2
Arginine	5.6	4.6	4.8	5.6	6.1	5.3	5.6	6.2	7.6
Glutamic acid	10.9	11.7	11.3	15.4	10.4	11.7	13.9	12.6	19.9
Aspartic acid	11	7.5	7.5	9.4	7.7	8.8	10.4	9.1	14.1
Proline	6.6	3.3	6.8	2.9	5.6	6.2	5.2	4.2	6.0
Glycine	5.7	4.2	4.9	4.8	5.2	5.9	4.8	6.4	4.5
Alanine	7.7	5.8	7.3	4.6	8.8	9.5	5.8	6.3	4.5

Table 2.
Amino acid composition (g/16 g nitrogen) of insect meals versus, soybean meal, and fish meal.

particularly n-6 PUFAs, but deficient in essential omega-3 fatty acids like the long chain eicosapentaenoic acid (EPA) and docosahexaenoic acid (DHA), compared to fish oil [22, 49]. This limitation restricts their use as a sole oil source in aquaculture feeds, as salmonids, for instance, require dietary EPA and DHA for optimal growth and development. The lipid content and fatty acid profile of insects can be influenced by factors such as diet and rearing conditions [27]. Commercially reared insects tend to have higher fat content and a different fatty acid profile compared to wild-caught insects. For instance, they often exhibit higher levels of linoleic acid and lower levels of linolenic acid due to their grain-based diets (**Table 3**) [53].

2.3 Carbohydrates

Generally, most of the insect meals used in aquafeeds exhibit low carbohydrate levels, typically below 20% [15]. Chitin, a polysaccharide constituting 5–20% of an insect's dry weight, forms a significant portion of the insect exoskeleton. It contributes 6.71–15.98% of the total carbohydrate content. While chitin is the second most abundant carbohydrate in biomass after cellulose, its impermeability in liquid media limits its digestibility. Insect meals contain varying amounts of chitin, ranging from negligible levels to over 10% of dry matter [54]. While most fish species cannot digest chitin, some species may be negatively affected by its presence. The limited digestibility of chitin is often cited as a constraint in the use of insect meals in aquaculture feeds [36]. However, on the other hand, published results indicate chitin and chitosan to be a promising immunostimulant for aquaculture, besides having other beneficial attributes [55]; other results indicate that dietary intake of chitosan (derivative of chitin) enhances the innate immune system and survivability of common carp (*Cyprinus carpio*) in ponds [55]. Survival of Nile tilapia also improved when they were fed with diets with black soldier fly larvae meal included [32], which was attributed to the chitin present in the black soldier fly larvae meal which is known to offer other beneficial effects to the fish such as boosting the immunity of the fish [56].

2.4 Minerals and vitamins

Minerals are categorized as macro-minerals (calcium, phosphorus, magnesium, sodium, potassium, and chloride) and micro-minerals (iron, zinc, copper, manganese, iodine, and selenium) based on their dietary requirements. Insects are a source of various minerals, including potassium, calcium, iron, magnesium, zinc, and selenium [45]. However, their calcium and phosphorus levels are generally lower than those in fishmeal, with the exception of calcium in black soldier fly larvae and phosphorus in houseflies [27]. Although most insect species utilized in fish feeds have low ash content, black soldier fly larvae stand out with a higher ash content and calcium concentration (7.6% of dry matter) [4]. While most insect exoskeletons primarily consist of protein and chitin, black soldier fly larvae possess a mineralized exoskeleton incorporating calcium and other minerals into their cuticle [57]. Insects, lacking a mineralized skeleton, generally exhibit low calcium levels, typically below 0.3% of dry matter, with exceptions like stoneflies and certain other species [42, 46]. With the exception of black soldier fly larvae, which has a calcium-to-phosphorus ratio of 8.1, the other insect species exhibit calcium-to-phosphorus ratios ranging from 0.2 to 1.2. These ratios are lower than the recommended levels of 1.1 to 1.4 for fish [36, 38]. To compensate for the sub-optimal calcium-to-phosphorus ratios in most insect meals, supplemental calcium should be added to aquaculture feeds when replacing fishmeal.

Nutritional components	Silkworm pupae	Black soldier fly larvae[b]	Housefly maggot	Yellow Mealworm	Lesser mealworm	House cricket[a]	Banded cricket[a]	Jamaican field cricket[a]	Fish oil	Soybean oil
Saturated fatty acids (%)										
Lauric, 12:0		21.4 [49.3] (42.6)	—	0.5	0.1	—	0.1	—	—	—
Myristic, 14:0		2.9 [6.8] (6.9)	5.5	4	1.4	0.7	1.65	0.43–0.76	3.4	—
Palmitic, 16:0	24.9	16.1 [10.5] (11.1)	31.1	21.1	26.4	23.4	23.5	20.75–24.54	16.1	10.6
Stearic, 18:0	5.4	5.7 [2.78] (1.3)	3.4	2.7	10.9	9.8	7.35	5.02–6.93	4.6	3.8
Monosaturated fatty acids (%)										
Palmitoleic, 16:1n-7	0.8	[3.5]	13.4	4	1.1	1.3	3.78	0.97–2.08	0.8	—
Oleic, 18: 1n-9	24.3	32.1 [11.8] (12.3)	24.8	37.7	35.9	23.8	29.14	28.65–30.98	12.7	21.8
Polyunsaturated fatty acids (%)										
Linoleic, 18:2n-6	6.3	4.5 [3.7] (3.6)	19.8	27.4	—	38	29.78	31.79–37.72	1.4	53.7
Eicosapentaenoic, 20:5n-3	0.2	0.03 [0] (1.66)	—	—	—	—	—	1.16	11.1	—
Docosahexaenoic, 22:6n-3	—	0.006 [0] (0.59)	—	—	—	—	0.07	0.15	29.1	—

[a] *Data from whole body meal. –: No reported data are available or there is no this component in the species.*

[b] *Values using cow manure as substrate. Round parentheses are the values obtained on using 50% of cow manure and 50% of fish offal as substrate. Square parentheses are values obtained on swine manure as substrate.*

Table 3.
Fatty acid composition of insect oil (total lipids), soya oil, and fish oil.

Most insects provide adequate magnesium to meet dietary requirements with black soldier fly larvae standing out with significantly higher magnesium concentrations [57]. Limited research is available on the sodium and potassium content of insects [42]. Additionally, insects are generally rich in trace minerals such as iron, zinc, copper, manganese, and selenium. Interestingly, iron and zinc levels in insects often correlate, with higher iron concentrations associated with higher zinc levels [58]. While insects provide several vitamins [59], their vitamin and mineral profiles are significantly influenced by their diet [37]. Sangha et al. [37] demonstrated that silkworm pupae are rich in vitamins with vitamin C being the most abundant vitamin followed by thiamine and niacin.

3. Insect species as aquafeeds

Insects are gaining attention as an environmentally sustainable and protein-rich alternative for aquaculture feed. In particular, insect meal is being explored as a substitute for fishmeal in aquaculture diets. The European Union's recent approval of insect meal from seven insect species for use in aquafeeds has spurred growing interest in this alternative [4, 60]. Following the principles of the circular economy, insects are seen as a promising and efficient ingredient for aquaculture feeding. As a result, many countries are increasingly turning to insect meals to replace fishmeal in their aquaculture industries. Due to its superior protein content and well-balanced amino acid profile, insect meal has emerged as a preferred alternative to fishmeal, providing a novel protein source for both aquatic feeds [39]. Insects represent a promising alternative to traditional animal-based protein sources for animal feed [15]. The European Commission in 2017 lifted its ban on the use of processed animal proteins derived from insects in aquafeeds under regulation EU-2017/893, making insect meal a viable option for fish farming. This regulation permits the use of seven specific insect species: black soldier fly (*H. illucens*), common housefly (*M. domestica*), yellow mealworm (*T. molitor*), lesser mealworm (*Alphitobius diaperinus*), house cricket (*A. domesticus*), banded cricket (*G. sigillatus*), and field cricket (*G. assimilis*). Also, regulation (EU) No 2019/1981 introduced a list of third countries that were authorized to export insect products complying with the mentioned Regulation (EU) No 2017/893. Further, many other countries have also developed national standards for the use of insect meals in the formulation of diets for animals including fish. In the United States of America, black soldier fly larvae, including dried whole larvae (since 2016) and black soldier fly meal (since 2018), is permitted for use in feed for aquaculture for salmonids such as salmon, trout, and char. In Canada, black soldier fly products have been authorized to feed broiler chickens, salmonids, tilapia and poultry including chickens, ducks, geese, and turkey [61]. In Australia and New Zealand, insects may be used as feed for aquaculture and many other countries in Asia and Africa.

Many of these insect species possess advantageous traits for aquaculture feed production, such as high feed conversion efficiency, short life cycles, and the capacity to thrive on various organic waste streams (substrates). Among these, black soldier fly and housefly have garnered attention in recent years due to their numerous advantages as a sustainable protein source in aquaculture [62]. Black soldier fly and yellow mealworms are currently the most widely studied insect species for aquaculture feed formulations [23]. Various researchers have conducted extensive investigations into the dietary use of black soldier fly and yellow mealworms meals in marine and freshwater carnivorous fish species, yielding promising outcomes [21]. Despite variations in propagation and production methods, insects have demonstrated significant potential as protein and oil sources in aquafeed. Numerous studies have shown that insect meals and oils can effectively replace traditional fishmeal and soybean meal in aquaculture diets [38, 39]. Therefore, studies have explored the efficacy of replacing fishmeal with insect meal in various fish diets, with varying results.

3.1 Black soldier fly (*Hermetia illucens*)

Black soldier fly larvae (**Figure 1**) have emerged as a highly sustainable option for aquafeed. Black soldier fly larvae meal exhibits a high protein content with an amino acid profile comparable to fishmeal, making it a well-balanced feed [63]. Black soldier fly, widely distributed in warm and temperate regions, is a promising protein source due to its short life cycle, ease of breeding, and ability to thrive on diverse organic waste materials. The fly exhibits rapid growth rates, high fertility, and the ability to convert waste into high-quality protein [64]. Due to their diminutive size, insects require significantly less space for breeding and farming. Moreover, insects have been reported to emit significantly lower levels of ammonia compared to domestic livestock [65]. A study conducted by Oonincx et al. [66] revealed that insects emit 80 times less ammonia than cattle on a weight-for-weight basis. This finding is particularly significant considering the 25 times greater impact of methane on global temperature compared to carbon dioxide. Wang & Shelomi [67] highlighted the advantage of insect farming, specifically black soldier fly larvae, in terms of reduced space requirements.

3.1.1 Chemical composition

Black soldier fly larvae are a valuable nutrient source, containing protein, lipids, minerals, vitamins, amino acids, and fatty acids. The black soldier fly larvae contain a moderate level of protein (31–59%) with an amino acid profile that closely resembles fishmeal and surpasses that of soybean meal [15, 27, 39]. The protein content of black soldier fly larvae is suitable for various cultured fish and crustacean species with moderate crude protein requirements. Additionally, black soldier fly larvae are a good

Figure 1.
Black soldier fly larvae.

source of lipids (11–49%) [27, 39]. The lipid profile is dominated by saturated fatty acids, particularly lauric acid [31, 68]. Black soldier fly larvae also contain minerals, particularly calcium [27, 42], and their fatty acid composition is significantly influenced by their dietary intake.

3.1.2 Nutritional value in fish

Numerous studies have demonstrated the potential of black soldier fly larvae to replace conventional fishmeal and soybean meal in aquaculture feeds without negatively impacting fish growth, feed efficiency, digestion, or fillet quality [69]. While partial replacement is commonly recommended, recent studies have indicated that complete replacement may be viable, especially for carnivorous fish species [69]. In yellow catfish, black soldier fly larvae meal can substitute up to 20% of fishmeal in traditional diets without significantly compromising growth performance [70]. A study conducted on rainbow trout (*Oncorhynchus mykiss*) evaluated the potential of a partially defatted black soldier fly larvae meal as a feed ingredient. The results indicated that a defatted black soldier fly larvae meal can be incorporated into trout diets up to a maximum of 40% without compromising growth, survival rate, condition factor, somatic indexes, fillet quality parameters, or intestinal morphology [71]. Research has demonstrated that defatted black soldier fly larvae meal can replace up to 50% of dietary fishmeal without adverse effects [32, 56, 72]. When the replacement level approached 75%, signs of dietary stress and intestinal histological damage became evident. Dietary substitution of black soldier fly larvae meal for fishmeal had a negligible effect on whole-body protein, lipid, or amino acid composition, while significantly influencing whole-body fatty acid composition. In general, these studies revealed that replacing fishmeal with black soldier fly larvae meal in the diets of sea-water Atlantic salmon and Nile tilapia is feasible without compromising growth, feed utilization, nutrient digestibility, liver characteristics, or fillet sensory qualities [32, 56, 73]. Feeding experiments conducted on European sea bass (*Dicentrarchus labrax*) revealed that partial replacement of fishmeal with defatted black soldier fly larvae meal, up to 50%, did not significantly affect growth performance. However, it resulted in a 15.6% reduction in feeding costs compared to the fishmeal-control diet [74].

3.2 Yellow mealworm (*Tenebrio molitor*)

Yellow mealworm has been evaluated as a potential alternative protein source to fishmeal in the diets of diverse fish species. The yellow mealworm, a member of the Tenebrionidae family, is one of the most promising insect species for mass production owing to its ease of breeding and feeding [75]. Mealworm larvae exhibit robust growth when fed on plant by-product diets [76] and possess a relatively short life cycle, with egg, larval, and pupal stages lasting 3–9 days, 26–76 days, and 5–17 days, respectively [77]. The larvae of yellow mealworms can be readily reared on low-nutritive plant materials and efficiently converts food waste and agricultural by-products into high-quality biomass [78]. These insects are commonly used as fish feed, either live, canned, dried, or powdered [79].

3.2.1 Chemical composition

The larval and pupal stages of yellow mealworms possess significant levels of proteins and lipids, including a range of essential amino acids, particularly methionine, as well as lipids and essential fatty acids. Mealworms exhibit a high protein content (44.1–60.3% DM) and lipid content (16.6–43.1% DM), possessing amino acid and fatty acid profiles suitable for inclusion in animal feeds [23]. Defatted yellow mealworms provide a substantial protein content of up to 63.84% and exhibit an amino acid composition comparable to fishmeal [80]. It has been observed that yellow mealworms larvae contain a higher proportion of monounsaturated fatty acids (MUFAs) compared to polyunsaturated fatty acids (PUFAs) and saturated fatty acids (SFAs) [81]. Langston et al. [82] found that yellow mealworms larvae reared on diverse substrates were rich in most minerals, with the exception of calcium. Furthermore, yellow mealworms contain a variety of physiologically active compounds, including anti-tumoral, antibacterial, antioxidant, and immunomodulatory substances [83].

3.2.2 Nutritional value in fish

Although studies have indicated that 25–30% inclusion of yellow mealworms in fish diets is optimal [84], recent research on rainbow trout has demonstrated improved performance with higher levels of fishmeal/yellow mealworm meal replacement [21]. Furthermore, yellow mealworm larvae have exhibited the highest apparent digestibility coefficient among the four insect meals tested in Nile tilapia [85], confirming their suitability as a protein source in fish diets. In rainbow trout diets, replacing up to 50% of fishmeal with yellow mealworm meal improved feed efficiency and protein utilization without affecting growth [75]. However, a 50% replacement with full-fat yellow mealworm meal in European seabass reduced growth compared to fishmeal-based diets [86]. The findings of Piccolo et al. [87] demonstrated the feasibility of substituting up to 25% of fishmeal protein in diets for juvenile gilthead sea bream without adverse effects on growth performance or whole-body proximate composition. A study by Su et al. [88] revealed that incorporation of 18% yellow mealworm meal into the diet of yellow catfish significantly enhanced the immune response and bacterial resistance without compromising growth performance. Feeding trials conducted on gilthead sea bream revealed that the inclusion of mealworm meal in the diet significantly impacted the free amino acid profiles of both the feed and fish muscle (**Figure 2**) [89].

3.3 Housefly (*Musca domestica*)

Houseflies represent the most diverse group of flies, capable of transforming decaying organic matter into a nutritionally rich animal-derived food source [90]. Both the larval (maggot) and adult stages of *Musca domestica* feed on manure and decaying organic waste. These insects are relatively easy to produce and process [91] and are generally more cost-effective than other animal protein sources. Housefly maggots can thrive on a wide range of organic wastes, including pig dung [92], cattle blood and wheat bran [93], cattle gut and rumen content [91], and poultry manure. The ability of housefly maggots to thrive on a wide range of substrates makes them a potential resource for transforming waste into valuable biomass rich in protein and fat [27].

3.3.1 Chemical composition

Housefly maggots, also known as maggot meal, are a source of protein and lipids. The protein content of housefly maggots ranges from 40 to 60%, while the lipid

Figure 2.
Yellow mealworms (Adapted from Makkar et al., 2014).

content exhibits even greater variability, ranging from 9 to 26%. Older larvae tend to exhibit lower protein content and higher lipid content [93]. Drying methods, such as sun-drying and oven-drying, can influence the nutrient composition, with sun-drying potentially leading to lower protein and higher lipid content [93]. While housefly maggots contain a relatively low level of crude fiber (usually less than 9%), they may have higher levels of acid detergent fiber. Although the phosphorus content of housefly maggots is comparable to that of black soldier fly larvae, calcium levels are approximately 15 times lower [27]. Housefly maggots, similar to black soldier fly larvae, exhibit a high lysine content, ranging from 5 to 8.2 g/100 g CP, with an average of 6.1 g/100 g CP. The fatty acid profile of housefly maggots is significantly influenced by substrate composition, with changes in fatty acid composition being one of the earliest observable responses to alterations in substrate [27]. Notably, the palmitoleic acid level in housefly maggots, at 17.1%, is considerably higher (4–15 times) than that observed in mealworms and house crickets. Conversely, the linoleic acid level in housefly maggots is significantly lower compared to these other insect species [27].

3.3.2 Nutritional value in fish

The application of house fly maggot meal as a fish food supplement has primarily been investigated in tilapia and catfish species, as well as other aquaculture species. Feeding trials have generally yielded encouraging results in catfish species. However, it is important to note that maggot meal inclusion should not exceed 30%, as higher inclusion rates tend to negatively impact growth performance [94]. A study investigated the cost-effectiveness and production of housefly maggots as a fish feed ingredient. Results showed that maggots could replace up to 60–70% of fishmeal in catfish (*Clarias gariepinus*) diets without negatively impacting growth and nutrient utilization [95]. A study was conducted to evaluate the efficacy of housefly maggots, cultured on poultry waste, as a replacement for fishmeal in aquafeeds at various inclusion levels (0, 25, 50, 75, and 100%). The study indicated that an inclusion level of 75% wet maggots in commercial fish feed is optimal for ensuring adequate utilization by catfish juveniles [96].

3.4 Silkworm (*Bombyx mori*)

Silkworms are the larval stage of silk-producing moth species. Approximately 90% of global silk production originates from the cocoons of the domestic mulberry silkworm (*Bombyx mori*), a moth belonging to the Bombycidae family. The life cycle of the silkworm caterpillar commences with the oviposition of eggs by adult moths [97]. Upon hatching, the caterpillars undergo a period of continuous feeding on mulberry and shea butter leaves, reaching a maximum size of approximately 10 centimeters within 4–6 weeks. Subsequently, the silkworm enters the pupal stage, constructing a protective cocoon from raw silk. The pupae eventually release an enzyme that induces an opening in the cocoon, allowing the emergence of adult moths [97]. To obtain silk fibers, the pupae are subjected to a killing process, such as drying, moist heating, or immersion in sodium hydroxide, prior to enzyme release [98]. Silkworm pupae are generated as a by-product following the extraction of silk from cocoons through spinning or reeling processes. Several silkworm species, including *Antheraea assamensis*, *Antheraea mylitta*, *Antheraea paphia*, and *Samia cynthia ricinii* [27], are collectively referred to as "silkworms." Nevertheless, the domestic mulberry silkworm (*Bombyx mori*) remains the most significant species, accounting for approximately 90% of global silk production.

3.4.1 Chemical composition

Silkworm pupae meal is a protein-rich feed ingredient characterized by high nutritional value. Its crude protein content ranges from 52 to 72%, with defatted meal exhibiting even higher levels [27]. Similar to other insects, silkworm pupae meal is relatively low in calcium and has a low calcium-to-phosphorus ratio. The lysine content of silkworm pupae meal, ranging from 6 to 7% in 100 g CP, and the methionine plus cystine levels, approximately 4%, are notably high. Conversely, the chitin content of pupae meal is relatively low, at around 3–4% DM [99]. Recent study from Kenya showed that the silkworm pupae meal is also rich in minerals and vitamins and that the overall nutrient content of the meal is influenced by the mulberry variety that the larvae consume [37].

3.4.2 Nutritional value in fish

Numerous studies have demonstrated the benefits of incorporating silkworm pupae meal into the diets of fish larvae and juveniles, including cyprinids and various other aquaculture species. Unlike other insect meals, silkworm pupae meal, regardless of whether it is defatted or not, has consistently yielded positive results in fish feeding trials. Moreover, the fat derived from silkworm pupae is considered beneficial [27, 100]. Additionally, feeding trials have revealed that both defatted and non-defatted silkworm pupae meals exhibit high digestibility in tilapia [101] and catfish [102]. The digestibility of both forms of silkworm pupae meal in cyprinids has been observed to surpass that of fishmeal [102]. In mirror carp (*Cyprinus carpio*), an 11-week feeding trial demonstrated that silkworm pupae meal is a promising sustainable functional feed component. The inclusion of silkworm pupae meal in carp diets resulted in improved growth performance and specific physiological parameters [103]. In rainbow shark (*Epalzeorhynchos frenatum*), feeding experiments revealed that silkworm pupae meal could effectively replace up to 30% of fishmeal in the diet [104]. Collectively, these findings suggest that silkworm pupae meal is a suitable protein source for partial replacement of fishmeal in aquaculture feeds (**Figure 3**) [102].

Figure 3.
Silkworm pupae.

3.5 Crickets

The house cricket (*Acheta domesticus*) is an excellent source of protein for both animals and human beings. Crickets offer a high economical protein source [105] and thus a sustainable solution to protein deficiency in animal feeds. Cricket farming is a feasible venture and has demonstrated success in some countries such as Thailand [106]. Cricket rearing requires simple locally available materials such as egg trays, which is important if a local community will adopt rearing of crickets in an easy affordable way. Currently, cricket farming is still under small-scale production in most parts of the world, but people are slowly adopting to cricket rearing largely for cash at the local markets, and this is why there is increased interest in cricket rearing. Cricket rearing has shown to be economical in water and feed consumption, since they consume little water and food. Cricket rearing is also time efficient and an environmentally safe way of alternative protein generation as they produce less greenhouse gases [105]. Crickets have a high feed conversion rate, thus providing high quality nutrients, which in turn are cheaper, efficient, and environmentally friendly protein source in comparison with other protein sources, which are expensive to maintain [107]. Crickets should thus be used as an alternative source of protein for animal feeds especially in developing countries.

3.5.1 Chemical composition

The house cricket also exhibits a high protein content, ranging from 55 to 67%. Both calcium and phosphorus levels in house crickets are higher than those observed in locusts or grasshoppers. Both lysine and methionine plus cystine levels in house crickets are lower than those observed in locust meal. The palmitoleic acid level in house crickets is approximately 15-fold lower than that found in housefly maggots and 4-fold lower than in mealworms [27]. Conversely, the linoleic acid level is higher in-house crickets. Regarding other insect species, field crickets (*Gryllus testaceus*) and Mormon crickets (*Anabrus simplex*) exhibit high protein content, approximately 60%.

While both species contain 10–13% lipids, the calcium content in Mormon crickets is relatively low at 2 mg/kg DM. Lysine levels are lower in field crickets compared to Mormon crickets, whereas the level of sulfur-containing amino acids (methionine plus cystine) is higher in field crickets.

3.5.2 Nutritional value in fish

The house cricket meal has been employed as a protein source in livestock feed, while its utilization in fish feed is still in its nascent stages [54]. Feeding trials have been conducted on a limited number of fish species. In hybrid tilapia (*Oreochromis* sp.), the impact of house cricket meal on growth performance was investigated. A feeding trial incorporating a diet containing 60% house cricket meal and 40% rice bran yielded the most favorable results in terms of survival and growth rate [108]. In perch (*Perca fluviatilis*), a 12-week feeding trial was conducted, involving the replacement of 25% of fishmeal with a mixture of house cricket meal. The study revealed no significant differences in survival rates between the control and experimental groups. However, a decrease in fish growth and an increase in feed conversion ratio (FCR) were observed in the experimental group. Compared to fish fillets from the control group, feeding with insect pellets resulted in a considerable elevation in linoleic acid levels and total n-6 fatty acid content in fish fillets. Nevertheless, the overall changes in fatty acid composition were minimal, and the nutritional value of fish fed a diet supplemented with house cricket meal remained unaffected [109]. These findings indicate that house cricket is a promising insect species for partial replacement of fishmeal in aquaculture feeds.

3.6 Locusts and grasshoppers

Locusts are type of grasshoppers that belong to the family Acrididae, the family comprises mainly grasshoppers with short antennas. They include migratory locust (*Migratoria locusta*), desert locust (*Schistocerca gregaria*), red locust (*Nomadacris septmfasciata*), and the brown locust (*Locusta pardalina*) [65].

3.6.1 Chemical composition

Locusts and other Orthoptera species are generally characterized by a high protein content, ranging from 50 to 65%, although lower values have been reported [27]. According to Kinyuru [110], the proximate composition of desert locusts was 52.3% protein, 12.0% crude fat, 19.0% crude fiber, 10.0% ash, 2.086 g/kg calcium, and iron content of 0.0483 g/kg on the dry matter basis. This protein value exhibited was higher compared to soybean meal (44.0% DM), sunflower cake (32.0% DM), cotton seed cake (40.9% DM), and sesame seed cake (43.8% DM). Calcium content, however, is relatively low, as observed in other insect species. The calcium and phosphorus content, at 1.3 and 1.1 mg/kg DM, respectively, is significantly lower compared to other insects such as black soldier fly meal, maggot meal, and housefly meal. Despite this, the calcium-to-phosphorus ratio is higher, primarily due to a lower phosphorus level. The essential amino acid profile is reasonably favorable. While the lysine level is lower than that of other insects, such as black soldier fly larvae, maggot, and housefly meals, the cystine plus methionine levels are relatively higher [27].

3.6.2 Nutritional value in fish

Desert locust meal has the potential to substitute up to 25% of dietary protein in catfish juveniles without significantly compromising growth performance. The presence of chitin may contribute to reduced performance and feed efficiency at higher substitution rates [111]. Meal derived from adult variegated grasshoppers (*Zonocerus variegatus*) can replace up to 25% of fishmeal (on a weight basis) in the diets of catfish fingerlings without any adverse impact on growth and nutrient utilization, provided that the dietary protein level remains constant. However, higher inclusion rates have been associated with decreased digestibility and performance [112]. A 91-day feeding trial utilizing a diet containing dried grasshopper meal revealed no significant alterations in hematological parameters. However, the study observed minor shrinkage in gill tissue and a reduction in ovarian steroidogenesis, which may potentially compromise fertility [113]. Migratory locust meal (*Locusta migratoria*) has demonstrated the potential to replace up to 25% of fishmeal in isoproteic diets for Nile tilapia fingerlings without any adverse impact on nutrient digestibility, growth performance, or hematological parameters [114].

4. Comparing the nutrient profiles of insects to traditional fish feed sources

Insects play a crucial role as a natural food source for fish, especially omnivorous and carnivorous species. These fish have high protein requirements [27, 49]. Sanchez-Muros et al. [36] conducted a detailed analysis of the nutritional composition of three different insect species, examining factors such as crude protein, amino acids, fat content, fatty acid profiles, and mineral concentrations. These insects are characterized by relatively high crude protein content, ranging from 42.1 to 63.3%. While this crude protein content is lower than that of fishmeal, it is comparable to soybean meal [115]. Among the insects, adult silkworms, locusts, and crickets have lower crude protein levels compared to the larvae of black soldier flies and houseflies. To optimize fish growth, supplementing diets with synthetic amino acids may be necessary, depending on specific species requirements. Compared to soybean meal, silkworms, black soldier flies, and houseflies exhibit more favorable amino acid profiles, making them potential substitutes for fishmeal in aquaculture feeds [38]. Insect species, except silkworms, have lower sulfur amino acid levels than fishmeal. Threonine levels are relatively consistent across the seven insect species, with silkworms showing higher concentrations [38]. Tryptophan levels are generally lower in the six insect species, excluding silkworms and housefly maggot meal.

Lipid content in insects can vary considerably within a species, influenced by factors like developmental stage and diet [50]. Insect fat content is generally lower than that of fishmeal, with insects accumulating fat during development [36]. Various insect species exhibit a wide range of fat content, from approximately 8% in mature locusts to around 36% in mealworm larvae [49]. Insect meals and fish oil differ significantly in fatty acid composition. Fish oil is richer in omega-3 fatty acids compared to insect meals [27]. Insect meals typically contain higher levels of saturated fatty acids than unsaturated fatty acids, with unsaturated fatty acids comprising about 60–70% of the total fat. Compared to fish oil, these insect species contain lower levels of eicosapentaenoic acid (EPA, 20:5n-3) and docosahexaenoic acid (DHA,

22:6n-3), as noted by Hawkey et al. [116], but they are richer in polyunsaturated fatty acids (PUFAs), particularly n-6 PUFAs [49]. Due to their lower eicosapentaenoic and docosahexaenoic acids content, insect meals are primarily suitable as a secondary oil source in aquafeeds.

5. Nutritional requirements of fish

Nutrition is the primary cost driver in fish farming, with feed expenses making up over 60% of operational costs. A substantial portion of these feed costs is attributed to the high protein content required in fish diets. Aquaculture feeds are carefully formulated and balanced to provide optimal ratios of protein, carbohydrates, and lipids in order to meet the specific nutritional needs of fish species [117]. Fishmeal is preferred in aquafeeds since it offers high crude protein content, a balanced amino acid profile, particularly in essential amino acids like lysine, methionine, and tryptophan, which are often deficient in plant-based proteins. Additionally, fishmeal is highly digestible and palatable, leading to increased feed intake and nutrient utilization by fish. It is also free from anti-nutritional factors [118]. The optimal nutrient balance in fish feed must be tailored to the specific species and life stage. Given the controlled environment of intensive aquaculture systems, feeds must provide a complete nutritional profile, as fish have no access to supplementary nutrients from nature.

Balance of dietary energy plays a crucial role in establishing feeding standards for fish. Energy is derived from the metabolism of lipids, carbohydrates, and amino acids (proteins) [119]. In fish feeds, carbohydrates and lipids are the primary energy sources, as they are more cost-effective compared to proteins. Dietary energy levels typically influence feed intake, which in turn can affect growth and nutrient utilization. For example, studies have shown that when dietary energy is low, fish tend to increase feed intake in an effort to meet their energy requirements [120]. On the other hand, higher dietary energy levels often result in reduced feed intake, which may lead to a decreased intake of essential nutrients and hinder growth. Additionally, excessive energy intake has been linked to increased fat deposition in the body and a reduction in nutrient utilization efficiency. Fish-fed diets with a dietary energy level of 3.0 kcal/g exhibited the best feed conversion ratio, while those given diets with a dietary energy level of 3.2 kcal/g showed the highest protein efficiency ratio [120]. These discrepancies may be attributed to several factors, such as variations in temperature and differences in production systems.

Protein requirements are a crucial aspect of fish nutrition. Proteins, composed of amino acids, are essential for growth, tissue repair [121], and can even serve as an energy source when in excess [122]. Protein, being the most expensive feed component, significantly impacts operational expenses in fish farming. Consequently, using protein as an energy source is economically impractical. Protein requirements vary based on factors such as life stage, temperature, dietary energy content, feeding rate, protein digestibility, and protein quality [117]. Research suggests a minimum crude protein requirement of 40–43% for fish in the growth phase, while other studies indicate a range of 30–35% [123]. Protein levels in fish diets are also influenced by dietary energy content. Imbalances in the protein-to-energy ratio, particularly when non-protein energy is low, can lead to reduced growth and protein efficiency ratio [124].

Fish need a balanced supply of both essential and non-essential amino acids, rather than simply a high protein content [125]. Diets formulated based solely on protein levels may not adequately provide all essential amino acids [126]. An imbalance in essential amino acids (EAAs) can negatively impact growth performance and lead to increased nitrogenous waste discharge into the environment [125]. While over 200 amino acids exist in nature, only 10 are essential for fish. These EAAs, including methionine, arginine, threonine, tryptophan, histidine, isoleucine, lysine, leucine, valine, and phenylalanine and must be obtained from their diet since they cannot be synthesized by fish in adequate amounts [122]. Lysine and methionine are commonly the most limiting amino acids in fish diets.

Lipids are a crucial non-protein energy source for fish. Adequate lipid levels in the diet optimize the use of dietary protein for growth. Lipids provide a higher energy content than carbohydrates and are more readily digested [127]. In addition to serving as an energy source, lipids supply essential fatty acids and assist in the absorption of fat-soluble vitamins [127]. The primary source of lipids in fish feeds is polyunsaturated fatty acids, including those from fish oil [117]. Fish oil is predominantly made up of triglycerides along with some fat-soluble vitamins [117]. The recommended minimum lipid inclusion level in fish diets typically ranges from 10 to 12% [123], while commercial feeds often contain around 4–6% lipid content [119].

Calcium, phosphorus, potassium, magnesium, sodium, iron, sulfur, chlorine, copper, manganese, iodine, chromium, zinc, fluorine, cobalt, selenium, and molybdenum are essential macro- and micro-minerals for fish [128]. Phosphorus requirements are influenced by dietary composition and its form. Phytate-bound phosphorus in plant-based ingredients requires phytase enzymes for better digestibility [129]. Calcium and phosphorus together account for 70% of the mineral content in fish. Potassium is vital for acid—base balance, nerve function, and enzyme activity. Zinc is a crucial micronutrient, especially for growth and antioxidant responses in fish [130]. Zinc deficiency can hinder growth rates. Fish require dietary supplementation of vitamins as they either cannot synthesize them or produce them at a rate insufficient to meet their needs [117]. Sugars and starches can be used as cost-effective, non-protein energy source in fish [119]. While carbohydrates are not essential for fish growth, they play a protein-sparing role when included in appropriate amounts, allowing dietary protein to be used more efficiently for growth [131]. In addition to providing energy for fish, carbohydrates also serve as binding agents, enhancing the water stability of manufactured feed pellets. The digestible carbohydrate content typically ranges from 27 to 40% in commercial feeds, [128].

6. Insect meals feeding strategies in aquaculture

While there is growing interest in using insect meals in aquaculture feeds, limited research has been conducted to determine optimal inclusion levels [132, 133]. Existing studies have yielded conflicting results regarding the optimal incorporation levels of insect meals in fish diets, with factors such as fish species, growth stage, feed formulation, insect processing methods, and feeding duration influencing the outcomes. Recently, a meta-analysis highlighted the variability in maximum insect meal inclusion levels in fish feed, influenced by factors such as fish species, growth stage, feed formulation, and insect processing methods [31]. However, most studies suggest that 20–30% might be the maximum inclusion level without negative impacts. Additionally, the dietary base (plant-based or animal-based) may influence insect meal requirements.

Most studies have focused on replacing fishmeal with insect meals at various levels in fish diets to reduce reliance on fishmeal for sustainability and cost concerns. This has involved either combining different insect species [134] or using single insect species supplemented with amino acids to meet essential amino acid requirements [135]. However, Hoffmann et al. [134] demonstrated that the type of insect meal used in combination can significantly influence fish growth and feed utilization. It has also been demonstrated that when the insect meals are combined with fishmeal, the performance of fish improves [56]. Including chitin from insect meals in fish diets may offer positive effects by influencing the gut microbiota and enhancing the innate immune system, particularly when included in moderate amounts (between 25 and 50 mg/kg) [136]. However, higher inclusion levels of insect meal have generally shown detrimental effects in most species, which are thought to be linked to the elevated chitin content at these higher concentrations [71, 137]. Moreover, the negative effects observed in aquatic animals when insect meals are included in aquafeeds may be attributed to the lower fatty acid levels in these diets compared to traditional fishmeal-based diets [25]. Other studies have suggested that the adverse impacts of higher insect meal inclusion levels could be linked to the presence of non-protein nitrogen in certain insects, which may result in an overestimation of the protein content in the diet [138].

7. Palatability and digestibility of insect-based fish diets

Palatability influences feed intake among diets with varying compositions and sensory properties [20]. Overall, the findings on the use of insect meal in fish diets are inconsistent, largely due to variations in the nutritional quality of different insect meals, the experimental conditions, and the inclusion levels used [139]. The impact of insect meal on feed palatability remains uncertain, potentially varying with the specific insect species or fish species. While some studies suggest a negative impact, others indicate a positive effect. For instance, increasing dietary black soldier fly meal levels in turbot diets reduced feed intake due to decreased palatability [137]. The meal of houseflies (*Musca domestica*) is highly palatable and easily digestible for Nile tilapia [19]. Black soldier fly larvae can synthesize saturated and monounsaturated fatty acids, including lauric acid, which is prone to oxidation and may affect the flavor of the final product [140]. However, it remains unclear whether these issues are related to the inclusion level of insect meal or specific characteristics of the insect meal itself [141]. Self-feeding systems allow researchers to study how fish regulate food intake based on their preferences and the nutritional characteristics of the feed [142].

While the use of insects as a food source has gained significant attention, research on nutrient and energy digestibility from insect-based diets remains limited [142]. Factors such as insect species, inclusion level, life stage, and fish species can influence digestibility [38]. It is essential to characterize and determine the bioavailability of nutrients in food ingredients to formulate fish diets that meet the specific nutritional needs of different species [143]. The apparent digestibility coefficient is a common method to assess nutrient availability in feed ingredients. It measures the proportion of ingested nutrients that are absorbed and not excreted in feces [143]. Dry matter apparent digestibility coefficient offers a broad measure of overall digestibility, indicating the proportion of organic and inorganic matter that can be digested. This can vary considerably depending on factors such as the content of insoluble carbohydrates and minerals [40, 143]. The impact of insect meal on nutrient digestibility can be influenced by chitin content, although the extent of this effect depends on the inclusion level [87]. Chitin, the primary component of insect exoskeletons, binds to proteins, reducing both apparent and true nitrogen digestibility. Chitin, a non-protein nitrogen, can lower feed digestibility and hinder growth performance [137]. Basto et al. [143]

compared the digestibility of defatted and non-defatted yellow mealworm and black soldier fly larvae meals in European seabass. Defatted insect meals were found to be more digestible than their non-defatted counterparts. However, yellow mealworm meal exhibited higher dry matter digestibility (72.4–85.2%) compared to black soldier fly larvae meal (48.3–53.7%) in European seabass. In contrast, yellow mealworm meal showed lower digestibility (45.9%) in white shrimp (*Penaeus vannamei*) [144]. Gilthead sea bream also experienced reduced nutrient digestibility due to higher chitin content [87]. Marono et al. [145] proposed that chitin can reduce gut transit time and physically obstruct protein degradation by enzymes, thereby hindering nutrient digestibility.

8. Impact of insect meals on fish growth and health and consumer acceptability

Growth analysis is crucial for determining how efficiently fish convert ingested nutrients into growth. Several insect species, including black soldier fly [146], yellow mealworm [84], housefly [147], and cricket [148], have been evaluated for their impact on growth performance and feed utilization in aquaculture. Among these, black soldier fly is the most extensively studied insect. Fawole et al. [146] conducted a 60-day study to evaluate the effects of replacing fishmeal with black soldier fly larvae meal at 25, 50, and 75% levels on the growth, nutrient utilization, and health of African catfish. The study found that a 50% inclusion of black soldier fly larvae meal resulted in the highest final body weight, weight gain, and specific growth rate compared to other inclusion levels. Fish fed a diet containing 50% black soldier fly larvae meal exhibited improved feed conversion ratio, protein efficiency ratio, and protein productive value [146]. Similar results have been reported with Nile tilapia from Kenya [32, 56]. Belghit et al. [73] demonstrated that a complete replacement of fish meal with black soldier fly meal in Atlantic salmon diets was feasible without negatively impacting growth or nutrient digestibility. Yellow mealworm, the second most extensively researched insect in aquaculture after black soldier fly, shows promise as an alternative protein source for aquafeed. Rema et al. [149] found that increasing levels of defatted yellow mealworm in the diet improved growth and feed utilization in rainbow trout, demonstrating its potential to fully replace fishmeal. In contrast, a study by Iaconisi et al. [150] found no significant impact on growth or feed efficiency when mealworm was used to partially replace fishmeal at 25 and 50% for 131 days in blackspot seabream. Some studies have reported negative impacts on growth performance and feed utilization when using yellow mealworm in certain fish species [151, 152]. These results suggest a need for improved processing methods of the ingredient, as well as further research to optimize its use in aquaculture. Other insects, including the housefly [147], mopane worm [153], chironomid [154], and cricket [148], have also shown promise as protein sources for promoting fish growth. However, further research is needed to fully assess their potential.

The impact of insect-based diets on various immune-related parameters such as gut health, disease resistance, blood biochemistry, organ histology, and gene expression has been investigated in several aquaculture species. Additionally, replacing 75% of fishmeal with yellow mealworm in the diet of African catfish did not negatively impact their overall health [146]. In yellow catfish, a diet containing 18% yellow mealworm was shown to enhance the immune response and improve disease resistance against a bacterial challenge caused by *Edwardsiella ictalurid* [88]. Similarly, in juvenile mandarin fish, adding yellow mealworm to their diet boosted immune system function [84]. In Atlantic salmon, complete replacement of fishmeal with black soldier fly meal did not adversely affect liver histology or the expression of pro-inflammatory genes in the head kidney [73]. A study on juvenile Japanese seabass showed that incorporating BSF meal into their diet did not alter intestinal histomorphology [155]. Similarly, replacing fishmeal with black soldier fly larval meal in African catfish diets had no significant effect on blood biochemical parameters [32, 146]. Furthermore, replacing fishmeal with 25 and 50% black soldier fly in the diet of zebrafish did not result in any significant changes to gut histology, stress levels, or immune response [25]. In contrast, when fishmeal was substituted with yellow mealworm in the diet of juvenile Pacific white shrimp, it enhanced survival rates following exposure to the pathogenic bacteria *Vibrio parahaemolyticus* [156].

The widespread adoption of insect-based aquafeed will likely depend on the acceptance of both aquaculture producers and consumers. While limited research has explored public perception of insects as feed ingredients, the majority of aquatic animal product consumers have expressed positive views. These positive perceptions are often driven by factors such as perceived safety, sustainability considerations, and the availability of information about insect-based products [157]. Raising awareness and providing information about aquatic products raised on insect-based feeds are crucial for driving consumer acceptance and positive perception [158]. Baldi et al. [158] propose that reducing information asymmetry can help increase consumer acceptance. An Italian study revealed that men and younger consumers were more likely to accept aquatic products produced with insect-based feeds [158], suggesting that factors such as gender and age can influence consumer acceptance. The study also found that respondents with greater knowledge about the products exhibited higher acceptance rates compared to those with limited information. Similarly, an Australian study by Sogari et al. [159] observed that males were more open to accepting insect-based foods than females. Verbeke [160] noted that consumer perceptions of using insects in aquafeed are expected to change over time, influenced by factors such as culture, familiarity, and past experiences. This suggests that as consumers become more informed about insect-based diets for aquatic products, acceptance will likely increase. Moreover, cultural differences and beliefs are likely to shape these perceptions, though further research is needed to verify this.

9. Conclusion

Insect meals as innovative feed biomass for aquafeeds have been discussed here and more so in replacing fishmeal in aquafeeds. These insects possess high protein, fats, and calories, which make them excellent nutrient sources in aquafeeds. Many feeding trials with various aquaculture species have shown that these insect meals could successfully replace partial fishmeal in aquafeeds. Replacement rates of less than 30% are recommended in most studies. Higher rates of substitution for fishmeal, and 100% substitution, have been found to be technically possible in some aquaculture species. Use of mixed protein sources including insects' meals and fishmeal to feed fish has been shown to produce even better growth performance. However, using insects to replace fishmeal to feed farmed fish has some issues. One is the nutritional values of insects, which are different among species and stages of development within a species. Therefore, when designing feeds on an industrial-scale production of insects, it is essential to have constant substrates to feed insects. It is also important to carry out nutrient analysis of the insect meals before incorporation in formulations.

Regarding the composition of amino acids and the digestibility of proteins, black soldier fly larvae meal is the most similar to fishmeal and therefore the most studied for replacement of fishmeal in aquafeeds. It is certain that in the near future, insect farming for insect meals as a fish feed ingredient will affect aquaculture substantially and make aquaculture green, profitable, and sustainable.

Acknowledgements

The authors gratefully acknowledge the financial support for this work from the following organizations and agencies: Australian Centre for International Agricultural Research (ACIAR) (ProteinAfrica: LS/2020/154); Global Affairs Canada (BRAINS project: P011585); Novo Nordisk Foundation (RefIPro: NNF22SA0078466); IKEA Foundation (G-2204-02144); the Rockefeller Foundation (WAVE-IN: 2021 FOD 030); Bill and Melinda Gates Foundation (INV-032416); European Commission (NESTLER Project: 101060762 and INNOECOFOOD Project: 101136739); the Curt Bergfors Foundation Food Planet Prize Award; Norwegian Agency for Development Cooperation, Section for Research, Innovation, and Higher Education (grant number RAF-3058 KEN-18/0005, CAP-Africa); the Swedish International Development Cooperation Agency (Sida); the Swiss Agency for Development and Cooperation (SDC); the Australian Centre for International Agricultural Research (ACIAR); the Government of Norway; the German Federal Ministry for Economic Cooperation and Development (BMZ); and the Government of the Republic of Kenya. The funders had no role in the study design, data collection and analysis, decision to publish, or manuscript preparation. The views expressed herein do not necessarily reflect the official opinion of the donors.

Author details

Isaac Osuga[1]*, Catherine Maindi[2], Vincent Mwashi[1] and Chrysantus Tanga[3]

1 Department of Animal Sciences, College of Agriculture and Natural Resources, Jomo Kenyatta University of Agriculture and Technology, Nairobi, Kenya

2 Department of Agricultural Economics and Extension, School of Agriculture, University of Embu, Embu, Kenya

3 International Centre of Insect Physiology and Ecology, Nairobi, Kenya

*Address all correspondence to: isaac.osuga@jkuat.ac.ke

References

[1] Thurstan RH, Roberts CM. The past and future of fish consumption: Can supplies meet healthy eating recommendations? Marine Pollution Bulletin. 2014;**89**(1-2):5-11. DOI: 10.1016/j.marpolbul.2014.09.016

[2] FAO. Fishery and Aquaculture Statistics – Yearbook 2021. Rome: FAO Yearbook of Fishery and Aquaculture Statistics; 2024. DOI: 10.4060/cc9523en [Accessed: December 30, 2024]

[3] Stankus A. State of world aquaculture 2020 and regional reviews: FAO webinar series. In: FAO Aquaculture Newsletter. Rome (Italy): FAO; 2021. pp. 17-18

[4] Alfiko Y, Xie D, Astuti RT, Wong J, Wang L. Insects as a feed ingredient for fish culture: Status and trends. Aquaculture and Fisheries. 2022;**7**(2):166-178

[5] Naylor RL, Hardy RW, Buschmann AH, Bush SR, Cao L, Klinger DH, et al. A 20-year retrospective review of global aquaculture. Nature. 2021;**591**:551-563. DOI: 10.1038/s41586-021-03308-6

[6] Hodar AR, Vasava RJ, Mahavadiya DR, Joshi NH. Fish meal and fish oil replacement for aqua feed formulation by using alternative sources: A review. Journal of Experimental Zoology India. 2020;**23**(1):13-21

[7] Xiao X, Jin P, Zheng L, Cai M, Yu Z, Yu J, et al. Effects of black soldier fly (*Hermetia illucens*) larvae meal protein as a fishmeal replacement on the growth and immune index of yellow catfish (*Pelteobagrus fulvidraco*). Aquaculture Research. 2018;**49**(2018):1569-1577

[8] Hussain SM, Khurram F, Naeem A, Shah SZH, Sarker PK, Naeem E, et al. A review on the prospects and potentials of fishmeal replacement with different animal protein sources. International Aquatic Research. 2024;**16**(1):7-16. DOI: 10.22034/iar.2024.1996466.1517

[9] Dawood MA. Nutritional immunity of fish intestines: Important insights for sustainable aquaculture. Reviews in Aquaculture. 2021;**13**(1):642-663

[10] Mugwanya M, Dawood MA, Kimera F, Sewilam H. Anthropogenic temperature fluctuations and their effect on aquaculture: A comprehensive review. Aquaculture and Fisheries. 2022;**7**(3):223-243

[11] Cashion T, Le Manach F, Zeller D, Pauly D. Most fish destined for fishmeal production are food-grade fish. Fish and Fisheries. 2017;**18**(5):837-844. DOI: 10.1111/faf.12209

[12] Gephart JA, Henriksson PJG, Parker RWR, Shepon A, Gorospe KD, Bergman K, et al. Environmental performance of blue foods. Nature. 2021;**597**:360-366

[13] Rosenzweig C, Tubiello FN, Sandalow D, Benoit P, Hayek MN. Finding and fixing food system emissions: The double helix of science and policy. Environmental Research Letters. 2021;**16**:061002. DOI: 10.1088/1748-9326/ac0134

[14] Daniel N. A review on replacing fish meal in aqua feeds using plant protein sources. International Journal of Fisheries and Aquatic Studies. 2018;**6**(2):164-179

[15] Barroso FG, de Haro C, Sánchez-Muros MJ, Venegas E, Martínez-Sánchez A, Pérez-Bañón C. The potential

of various insect species for use as food for fish. Aquaculture. 2014;**422**:193-201

[16] Gaudioso G, Marzorati G, Faccenda F, Weil T, Lunelli F, Cardinaletti G, et al. Processed animal proteins from insect and poultry by-products in a fish meal-free diet for rainbow trout: Impact on intestinal microbiota and inflammatory markers. International Journal of Molecular Sciences. 2021;**22**(11):5454

[17] Aragão C, Gonçalves AT, Costas B, Azeredo R, Xavier MJ, Engrola S. Alternative proteins for fish diets: Implications beyond growth. Animals. 2022;**12**(9):1211. DOI: 10.3390/ani12091211

[18] Macusi ED, Cayacay MA, Borazon EQ, Sales AC, Habib A, Fadli N, et al. Protein fishmeal replacement in aquaculture: A systematic review and implications on growth and adoption viability. Sustainability. 2023;**15**(16):12500

[19] Wang YR, Wang L, Zhang CX, Song K. Effects of substituting fishmeal with soybean meal on growth performance and intestinal morphology in orange-spotted grouper (*Epinephelus coioides*). Aquaculture Reports. 2017;**5**:52-57

[20] Alves APDC, Paulino RR, Pereira RT, da Costa DV, e Rosa PV. Nile tilapia fed insect meal: Growth and innate immune response in different times under lipopolysaccharide challenge. Aquaculture Research. 2021;**52**(2):529-540

[21] Terova G, Gini E, Gasco L, Moroni F, Antonini M, Rimoldi S. Effects of full replacement of dietary fishmeal with insect meal from *Tenebrio molitor* on rainbow trout gut and skin microbiota. Journal of Animal Science and Biotechnology. 2021;**12**:1-14

[22] Gasco L, Acuti G, Bani P, Dalle Zotte A, Danieli PP, De Angelis A, et al. Insect and fish by-products as sustainable alternatives to conventional animal proteins in animal nutrition. Italian Journal of Animal Science. 2020;**19**(1):360-372

[23] Gasco L, Gai F, Maricchiolo G, Genovese L, Ragonese S, Bottari T, et al. Fishmeal alternative protein sources for aquaculture feeds. In: Feeds for the Aquaculture Sector: Current Situation and Alternative Sources. Gewerbestrasse (Switzerland): Springer Cham; 2018. pp. 1-28

[24] Veldkamp T, Meijer N, Alleweldt F, Deruytter D, Van Campenhout L, Gasco L, et al. Overcoming technical and market barriers to enable sustainable large-scale production and consumption of insect proteins in Europe: A susinchain perspective. Insects. 2022;**13**(3):281

[25] Zarantoniello M, Zimbelli A, Randazzo B, Compagni MD, Truzzi C, Antonucci M, et al. Black soldier Fly (*Hermetia illucens*) reared on roasted coffee by-product and *Schizochytrium* sp. as a sustainable terrestrial ingredient for aquafeeds production. Aquaculture. 2020;**518**:734659

[26] Randazzo B, Zarantoniello M, Cardinaletti G, Cerri R, Giorgini E, Belloni A, et al. *Hermetia illucens* and poultry by-product meals as alternatives to plant protein sources in gilthead seabream (*Sparus aurata*) diet: A multidisciplinary study on fish gut status. Animals. 2021;**11**(3):677

[27] Makkar HPS, Tran G, Heuzé V, Ankers P. State-of-the-art on use of insects as animal feed. Animal Feed Science and Technology. 2014;**197**:1-33. DOI: 10.1016/j.anifeedsci.2014.07.008

[28] Rumbos CI, Karapanagiotidis IT, Mente E, Athanassiou CG. The lesser

mealworm *Alphitobius diaperinus*: A noxious pest or a promising nutrient source. Reviews in Aquaculture. 2019;**11**(4):1418-1437

[29] Masson MV, de Souza Tavares W, Alves JM, Ferreira-Filho PJ, Barbosa LR, Wilcken CF, et al. Bioecological aspects of the common black field cricket, *Gryllus assimilis* (Orthoptera: Gryllidae) in the laboratory and in eucalyptus (Myrtaceae) plantations. Journal of Orthoptera Research. 2020;**29**(1):83-89

[30] de Souza-Vilela I, Andrew NR, Ruhnke I. Insect protein in animal nutrition. Animal Production Science. 2019;**59**:2029-2036

[31] Liland NS, Araujo P, Xu XX, Lock EJ, Radhakrishnan G, Prabhu AJP, et al. A meta-analysis on the nutritional value of insects in aquafeeds. Journal of Insects as Food and Feed. 2021;**7**(5):743-759

[32] Kariuki MW, Barwani DK, Mwashi V, Kioko JK, Munguti JM, Tanga CM, et al. Partial replacement of fishmeal with black soldier fly larvae meal in Nile tilapia diets improves performance and profitability in earthen pond. Scientific African. 2024;**24**:e02222

[33] Prabhu AJP, Schrama JW, Kaushik SJ. Mineral requirements of fish: A systematic review. Reviews in Aquaculture. 2014;**8**(2):172-219. DOI: 10.1111/raq.12090

[34] Hua K. A meta-analysis of the effects of replacing fish meals with insect meals on growth performance of fish. Aquaculture. 2021;**530**:735732

[35] Gasco L, Biasato I, Dabbou S, Schiavone A, Gai F. Animals fed insect-based diets: State-of-the-art on digestibility, performance and product quality. Animals. 2019;**9**(4):170

[36] Sanchez-Muros MJ, Barroso FG, Manzano-Agugliaro F. Insect meal as renewable source of food for animal feeding: A review. Journal of Cleaner Production. 2014;**65**:16-27

[37] Sangha MN, Barwani DK, Xavier C, Muhonja L, Moseti K, Karanja PN, et al. We are what we eat: Implications of host plant suitability on sustainable production of silkworm pupae as novel ingredient with dietary and health benefits. PLoS One. 2024;**19**(12):e0316290. DOI: 10.1371/journal.pone.0316290

[38] Henry M, Gasco L, Piccolo G, Fountoulaki E. Review on the use of insects in the diet of farmed fish: Past and future. Animal Feed Science and Technology. 2015;**203**:1-22

[39] Nogales-Mérida S, Gobbi P, Józefiak D, Mazurkiewicz J, Dudek K, Rawski M, et al. Insect meals in fish nutrition. Reviews in Aquaculture. 2019;**11**(4):1080-1103

[40] National Research Council. Nutrient Requirements of Fish and Shrimp. Washington: National Academies Press; 2011

[41] Kiron V. Fish immune system and its nutritional modulation for preventive health care. Animal Feed Science and Technology. 2012;**173**(1-2):111-133

[42] Finke MD. Complete nutrient content of four species of feeder insects. Zoo Biology. 2013;**32**(1):27-36

[43] Finke MD. Complete nutrient content of three species of wild caught insects, pallid-winged grasshopper, rhinoceros beetles and white-lined sphinx moth. Journal of Insects as Food and Feed. 2015;**1**(4):281-292

[44] Tran G, Heuzé V, Makkar HP. Insects in fish diets. Animal Frontiers. 2015;**5**(2):37-44

[45] Rumpold BA, Schlüter OK. Potential and challenges of insects as an innovative source for food and feed production. Innovative Food Science & Emerging Technologies. 2013;**17**:1-11

[46] Oonincx DGAB, Finke MD. Nutritional value of insects and ways to manipulate their composition. Journal of Insects as Food and Feed. 2021;7(5):639-659

[47] Bowyer JN, Qin JG, Stone DA. Protein, lipid and energy requirements of cultured marine fish in cold, temperate and warm water. Reviews in Aquaculture. 2013;**5**(1):10-32

[48] Henry M, Fountoulaki E. Optimal dietary protein/lipid ratio for improved immune status of a newly cultivated Mediterranean fish species, the shi drum *Umbrina cirrosa*, L. Fish & Shellfish Immunology. 2014;**37**(2):215-219

[49] Van Huis A. Insects as food and feed, a new emerging agricultural sector: A review. Journal of Insects as Food and Feed. 2020;**6**(1):27-44

[50] Barroso FG, Sánchez-Muros MJ, Rincón MÁ, Rodriguez-Rodriguez M, Fabrikov D, Morote E, et al. Production of n-3-rich insects by bioaccumulation of fishery waste. Journal of Food Composition and Analysis. 2019;**82**:103237

[51] Liu X, Chen X, Wang H, Yang Q, Ur Rehman K, Li W, et al. Dynamic changes of nutrient composition throughout the entire life cycle of black soldier fly. PLoS One. 2017;**12**(8):e0182601

[52] Tocher DR. Fatty acid requirements in ontogeny of marine and freshwater fish. Aquaculture Research. 2010;**41**(5):717-732

[53] Paul A, Frederich M, Megido RC, Alabi T, Malik P, Uyttenbroeck R, et al. Insect fatty acids: A comparison of lipids from three orthopterans and *Tenebrio molitor* L. larvae. Journal of Asia-Pacific Entomology. 2017;**20**(2):337-340

[54] Nikoletta H. Insects as animal feed. Magyar Allatorvosok Lapja. 2019;**141**(2):117-128

[55] Gopalakannan A, Arul V. Immunomodulatory effects of dietary intake of chitin, chitosan and levamisole on the immune system of *Cyprinus carpio* and control of *Aeromonas hydrophila* infection in ponds. Aquaculture. 2006;**255**(1-4):179-187. DOI: 10.1016/j.aquaculture.2006.01.012

[56] Wachira MN, Osuga IM, Munguti JM, Ambula MK, Subramanian S, Tanga CM. Efficiency and improved profitability of insect-based Aquafeeds for farming Nile tilapia fish (*Oreochromis niloticus* L.). Animals. 2021;**11**(9):2599. DOI: 10.3390/ani11092599

[57] Spranghers T, Ottoboni M, Klootwijk C, Ovyn A, Deboosere S, De Meulenaer B, et al. Nutritional composition of black soldier fly (*Hermetia illucens*) prepupae reared on different organic waste substrates. Journal of the Science of Food and Agriculture. 2017;**97**(8):2594-2600

[58] Mwangi MN, Oonincx DG, Stouten T, Veenenbos M, Melse-Boonstra A, Dicke M, et al. Insects as sources of iron and zinc in human nutrition. Nutrition Research Reviews. 2018;**31**(2):248-255

[59] Schabel HG. Forest insects as food: A global review. Forest Insects as Food: Humans Bite Back. 2010;**37**:64

[60] Tran HQ, Nguyen TT, Prokešová M, Gebauer T, Doan HV, Stejskal V. Systematic review and meta-analysis of

production performance of aquaculture species fed dietary insect meals. Reviews in Aquaculture. 2022;**14**(3):1637-1655

[61] Lähteenmäki-Uutela A, Marimuthu SB, Meijer N. Regulations on insects as food and feed: A global comparison. Journal of Insects as Food and Feed. 2021;7(5):849-856

[62] Lock EJ, Biancarosa I, Gasco L. Insects as raw materials in compound feed for aquaculture. In: Halloran A, Flore R, Vantomme P, Roos N, editors. Edible Insects in Sustainable Food Systems. Cham, Switzerland: Springer; 2018. pp. 263-276

[63] Cummins VC Jr, Rawles SD, Thompson KR, Velasquez A, Kobayashi Y, Hager J, et al. Evaluation of black soldier fly (*Hermetia illucens*) larvae meal as partial or total replacement of marine fish meal in practical diets for Pacific white shrimp (*Litopenaeus vannamei*). Aquaculture. 2017;**473**:337-344

[64] Smetana S, Schmitt E, Mathys A. Sustainable use of *Hermetia illucens* insect biomass for feed and food: Attributional and consequential life cycle assessment. Resources, Conservation and Recycling. 2019;**144**:285-296

[65] van Huis A, Itterbeeck JV, Klunder H, Mertens E, Halloran A, Muir G, et al. Edible Insects: Future Prospects for Food and Feed Security. 2013

[66] Oonincx DG, Van Itterbeeck J, Heetkamp MJ, Van Den Brand H, Van Loon JJ, Van Huis A. An exploration on greenhouse gas and ammonia production by insect species suitable for animal or human consumption. PLoS One. 2010;**5**(12):e14445

[67] Wang YS, Shelomi M. Review of black soldier fly (*Hermetia illucens*) as animal feed and human food. Food. 2017;**6**(10):91

[68] Sealey WM, Gaylord TG, Barrows FT, Tomberlin JK, McGuire MA, Ross C, et al. Sensory analysis of rainbow trout, *Oncorhynchus mykiss*, fed enriched black soldier fly prepupae, *Hermetia illucens*. Journal of the World Aquaculture Society. 2011;**42**(1):34-45

[69] Bruni L, Belghit I, Lock EJ, Secci G, Taiti C, Parisi G. Total replacement of dietary fish meal with black soldier fly (*Hermetia illucens*) larvae does not impair physical, chemical or volatile composition of farmed Atlantic salmon (*Salmo salar* L.). Journal of the Science of Food and Agriculture. 2020;**100**(3):1038-1047

[70] Hu J, Wang G, Huang Y, Sun Y, He F, Zhao H, et al. Effects of substitution of fish meal with black soldier fly (*Hermetia illucens*) larvae meal, in yellow catfish (*Pelteobagrus fulvidraco*) diets. Israeli Journal of Aquaculture-Bamidgeh. 2017;**69**(1):1382-1391

[71] Renna M, Schiavone A, Gai F, Dabbou S, Lussiana C, Malfatto V, et al. Evaluation of the suitability of a partially defatted black soldier fly (*Hermetia illucens* L.) larvae meal as ingredient for rainbow trout (*Oncorhynchus mykiss* Walbaum) diets. Journal of Animal Science and Biotechnology. 2017;**8**:1-13

[72] Li S, Ji H, Zhang B, Zhou J, Yu H. Defatted black soldier fly (*Hermetia illucens*) larvae meal in diets for juvenile Jian carp (*Cyprinus carpio* var. Jian): Growth performance, antioxidant enzyme activities, digestive enzyme activities, intestine and hepatopancreas histological structure. Aquaculture. 2017;**477**:62-70

[73] Belghit I, Lock EJ, Fumière O, Lecrenier MC, Renard P, Dieu M, et al.

Species-specific discrimination of insect meals for aquafeeds by direct comparison of tandem mass spectra. Animals. 2019;**9**(5):222

[74] Abdel-Tawwab M, Khalil RH, Metwally AA, Shakweer MS, Khallaf MA, Abdel-Latif HMR. Effects of black soldier fly (*Hermetia illucens* L.) larvae meal on growth performance, organs-somatic indices, body composition, and hemato-biochemical variables of European sea bass, *Dicentrarchus labrax*. Aquaculture. 2020;**522**:735136. DOI: 10.1016/j.aquaculture.2020.735136

[75] Belforti M, Gai F, Lussiana C, Renna M, Malfatto V, Rotolo L, et al. *Tenebrio molitor* meal in rainbow trout (*Oncorhynchus mykiss*) diets: Effects on animal performance, nutrient digestibility and chemical composition of fillets. Italian Journal of Animal Science. 2015;**14**(4):4170

[76] Li L, Zhao Z, Liu H. Feasibility of feeding yellow mealworm (*Tenebrio molitor* L.) in bioregenerative life support systems as a source of animal protein for humans. Acta Astronautica. 2013;**92**(1):103-109

[77] Adámková A, Adámek M, Mlček J, Borkovcová M, Bednářová M, Kouřimská L, et al. Welfare of the mealworm (*Tenebrio molitor*) breeding with regard to nutrition value and food safety. Potravinarstvo Slovak Journal of Food Sciences. 2017;**11**(1):460-465. DOI: 10.5219/779

[78] Tan SW, Lai KS, Loh JY. Effects of food wastes on yellow mealworm *Tenebrio molitor* larval nutritional profiles and growth performances. Examines in Marine Biology & Oceanography. 2018;**2**(1):173-178

[79] Veldkamp T, Van Duinkerken G, van Huis A, Lakemond CMM, Ottevanger E, Bosch G, et al. Insects as a Sustainable Feed Ingredient in Pig and Poultry Diets: A Feasibility Study= Insecten Als Duurzame Diervoedergrondstof in Varkens-en Pluimveevoeders: Een haalbaarheidsstudie (No. 638). Lelystad (Netherlands): Wageningen UR Livestock Research; 2012

[80] Zhang J, Dong Y, Song K, Wang L, Li X, Tan B, et al. Effects of the replacement of dietary fish meal with defatted yellow mealworm (*Tenebrio molitor*) on juvenile large yellow croakers (*Larimichthys crocea*) growth and gut health. Animals. 2022;**12**(19):2659

[81] Errico S, Spagnoletta A, Moliterni S, Dimatteo S, Verardi A, Sangiorgio P, et al. *Tenebrio molitor* as a new promising alternative in the production of feed and food. Environmental Engineering and Management Journal. 2022;**21**(10):1657-1671

[82] Langston K, Selaledi L, Yusuf A. Evaluation of alternative substrates for rearing the yellow mealworm *Tenebrio molitor* (L). International Journal of Tropical Insect Science. 2023;**43**(5):1523-1530

[83] Lee SJ, Yu JS, Parker BL, Skinner M, Je YH, Kim JS. Production of antibacterial *Bombyx mori cecropin* A in mealworm-pathogenic *Beauveria bassiana* ERL1170. Journal of Industrial Microbiology and Biotechnology. 2015;**42**(1):151-156

[84] Sankian Z, Khosravi S, Kim YO, Lee SM. Effects of dietary inclusion of yellow mealworm (*Tenebrio molitor*) meal on growth performance, feed utilization, body composition, plasma biochemical indices, selected immune parameters and antioxidant enzyme activities of mandarin fish (*Siniperca scherzeri*) juveniles. Aquaculture. 2018;**496**:79-87

[85] Fontes TV, de Oliveira KRB, Gomes Almeida IL, Orlando TM, Rodrigues PB, da Costa DV, et al. Digestibility of insect meals for Nile tilapia fingerlings. Animals. 2019;**9**(4):181

[86] Gasco L, Henry M, Piccolo G, Marono S, Gai F, Renna M, et al. *Tenebrio molitor* meal in diets for European sea bass (*Dicentrarchus labrax* L.) juveniles: Growth performance, whole body composition and in vivo apparent digestibility. Animal Feed Science and Technology. 2016;**220**:34-45

[87] Piccolo G, Marono S, Gasco L, Iannaccone F, Bovera F, Nizza A. Use of *Tenebrio molitor* larvae meal in diets for gilthead seabream *Sparus aurata juveniles*. In: Insects to Feed the World. Turin (Italy): University of Turin; 2014. pp. 68-68

[88] Su J, Gong Y, Cao S, Lu F, Han D, Liu H, et al. Effects of dietary *Tenebrio molitor* meal on the growth performance, immune response and disease resistance of yellow catfish (*Pelteobagrus fulvidraco*). Fish & Shellfish Immunology. 2017;**69**:59-66

[89] Iaconisi V, Secci G, Sabatino G, Piccolo G, Gasco L, Papini AM, et al. Effect of mealworm (*Tenebrio molitor* L.) larvae meal on amino acid composition of gilthead sea bream (*Sparus aurata* L.) and rainbow trout (*Oncorhynchus mykiss* W.) fillets. Aquaculture. 2019;**513**:734403

[90] Hussein M, Pillai VV, Goddard JM, Park HG, Kothapalli KS, Ross DA, et al. Sustainable production of housefly (*Musca domestica*) larvae as a protein-rich feed ingredient by utilizing cattle manure. PLoS One. 2017;**12**(2):e0171708

[91] Anene A, Afam-Anene OC, Ike K, Ekekwe NH. Preliminary investigations on quantity and proximate quality of maggots produced from four different sources of livestock wastes. Journal of Research in Biology. 2013;**3**(6):1060-1065

[92] Pastor BE, Elena H, Ková I, Kozánek MI, Martínez-Sánchez AN, Taká PE. Effect of the size of the pupae, adult diet, oviposition substrate and adult population density on egg production in *Musca domestica* (Diptera : Muscidae). European Journal of Entomology. 2011;**108**:587-596

[93] Aniebo AO, Owen OJ. Effects of age and method of drying on the proximate composition of housefly larvae (*Musca domestica* Linnaeus) meal (HFLM). Pakistan Journal of Nutrition. 2010;**9**(5):485-487

[94] Saleh HH. Effect of feeding on fresh (wet) housefly maggots (*Musca domestica*) with or without artificial diet on water quality and growth rates of African catfish (*Clarias gariepinus* Burchell, 1822) fry under laboratory conditions. Journal of Zoological Research. 2020;**2**(2):15-22

[95] Kolawole AA, Ugwumba AAA. Economic evaluation of different culture enclosures for *Musca domestica* larval production and their utilization for *Clarias gariepinus* (Burchell, 1822) fingerlings diets. Notulae Scientia Biologicae. 2018;**10**(4):466-474

[96] Ipinmoroti MO, Akanmu OA, Iyiola AO. Utilisation of house fly maggots (*Musca domestica*) as replacement for fish meal in the diets of Clarias gariepinus juveniles. Journal of Insects as Food and Feed. 2019;**5**(2):69-76

[97] Banday MT, Adil S, Sheikh IU, Hamadani H, Qadri FI, Sahfi ME, et al. The use of silkworm pupae (*Bombyx mori*) meal as an alternative protein source for poultry. World's Poultry Science Journal. 2023;**79**(1):119-134

[98] Jintasataporn O. Production performance of broiler chickens fed with silkworm pupa (*Bombyx mori*). Journal of Agricultural Science and Technology. A. 2012;**2**(4A):505

[99] Suresh HN, Mahalingam CA, Pallavi. Amount of chitin, chitosan and chitosan based on chitin weight in pure races of multivoltine and bivoltine silkworm pupae *Bombyx mori* L. International Journal of Science and Nature. 2012;**3**(1):214-216

[100] Sun Y, Chang AK, Wen Z, Li Y, Du X, Li S. Effect of replacing dietary fish meal with silkworm (*Bombyx mori* L) caterpillar meal on growth and non-specific immunity of sea cucumber A *Postichopus japonicus* (Selenka). Aquaculture Research. 2014;**45**(7):1246-1252

[101] Feng QF, Wang SW, Huang XZ, Li H, Liu CJ, Lu H, et al. Effects of oxidized silkworm (*Bombyx mori* L.) pupae on growth performance, and intestine, liver and muscle histology and function of *Gif Tilapia* (*Oreochromis niloticus*). Aquaculture Research. 2021;**52**(9):4127-4137

[102] Karthick Raja P, Aanand S, Stephen Sampathkumar J, Padmavathy P. Silkworm pupae meal as alternative source of protein in fish feed. Journal of Entomology and Zoology Studies. 2019;**7**(4):78-85

[103] Wan AH, Snellgrove DL, Davies SJ. A comparison between marine and terrestrial invertebrate meals for mirror carp (*Cyprinus carpio*) diets: Impact on growth, haematology and health. Aquaculture Research. 2017;**48**(9):5004-5016

[104] Raja PK, Aanand S, Sampathkumar JS, Padmavathy P. Effect of silkworm (*Bombyx mori*) pupae on the growth and maturation of rainbow shark *Epalzeorhynchos frenatum* (Fowler, 1934) under captive rearing. Indian Journal of Fisheries. 2020;**67**(4):89-96

[105] Caparros Megido R, Alabi T, Nieus C, Blecker C, Danthine S, Bogaert J, et al. Optimisation of a cheap and residential small-scale production of edible crickets with local by-products as an alternative protein-rich human food source in Ratanakiri Province, Cambodia. Journal of the Science of Food and Agriculture. 2016;**96**(2):627-632

[106] Hanboonsong Y, Jamjanya T, Durst PB. Six-legged Livestock: Edible Insect Farming, Collection and Marketing in Thailand. FAO, Regional Office for Asia and the Pacific Bangkok; 2013

[107] Payne CLR, Scarborough P, Rayner M, Nonaka K. Are edible insects more or less 'healthy' than commonly consumed meats? A comparison using two nutrient profiling models developed to combat over-and undernutrition. European Journal of Clinical Nutrition. 2016;**70**(3):285-291

[108] Lee SW, Tey HC, Wendy W, Zahari MW. The effect of house cricket (*Acheta domesticus*) meal on growth performance of red hybrid tilapia (*Oreochromis* sp.). International Journal of Aquatic Science. 2017;**8**(2):7882

[109] Tilami SK, Turek J, Červený D, Lepič P, Kozák P, Burkina V, et al. Insect meal as a partial replacement for fish meal in a formulated diet for perch *Perca fluviatilis*. Turkish Journal of Fisheries and Aquatic Sciences. 2020;**20**(12):867-878

[110] Kinyuru J, Kipkoech C, Imathiu S, Konyole S, Roos N. Acceptability of cereal-cricket porridge compared to cereal and cereal-milk- porridges among

caregivers and nursery school children in Uasin Gishu, Kenya. International Journal of Tropical Insect Science. 2021;**41**(3):2007-2013. DOI: 10.1007/s42690-020-00388-1

[111] Balogun BI, Oniye SJ, Auta J, Lakpini CAM, Obeke FO. Growth Performance and Feed Utilization of *Clarias gariepinus* (Teugels) Fed Different Dietary Levels of Soaked *Bauhinia monandra* (Kutz) Seed Meal. Minna (Nigeria): FISON; 2011. pp. 66-74

[112] Alegbeleye WO, Obasa SO, Olude OO, Otubu K, Jimoh W. Preliminary evaluation of the nutritive value of the variegated grasshopper (*Zonocerus variegatus* L.) for African catfish *Clarias gariepinus* (Burchell 1822) fingerlings. Aquaculture Research. 2012;**43**(3):412-420

[113] Johri RJ, Rakhi Singh RS, Johri PK. Effect of different formulated plant and animal diet on hematology of *Clarias batrachus* Linn. under laboratory conditions. Biochemical and Cellular Archives. 2010;**10**(2):283-291

[114] Abanikannda MF. Nutrient Digestibility and Haematology of Nile Tilapia (*Oreochromis niloticus*) Fed with Varying Levels of Locust (*Locusta migratoria*) Meal. Abeokuta, Ogun State: Bachelor of Aquaculture and Fisheries Management, Federal University of Agriculture; 2012

[115] Allegretti G, Schmidt V, Talamini E. Insects as feed: Species selection and their potential use in Brazilian poultry production. World's Poultry Science Journal. 2017;**73**(4):928-937

[116] Hawkey KJ, Lopez-Viso C, Brameld JM, Parr T, Salter AM. Insects: a potential source of protein and other nutrients for feed and food. Annual review of animal biosciences. Annual Reviews, Palo Alto (USA). 2021;**9**(1):333-354

[117] Geay F, Kestemont P. Feeding and nutrition of percid fishes during ongrowing stages. In: Biology and Culture of Percid Fishes: Principles and Practices. 2015. pp. 587-622

[118] Ginindza J. Effect Protein Levels on Nutrient and Energy Digestibility in Diet of *Arctic charr* (*Salvilinus alpinus*). Iceland: United Nations University Fisheries Training Programme; 2012. Final project

[119] Robinson EH, Li MH. A Brief Overview of Catfish Nutrition. USA: Mississippi Agricultural & Forestry Experimental Station. (Research report Vol. 24(16)), Mississippi State University (USA); 2015

[120] Aderolu AZ, Lawal MO, Awobajo FO, Olaniyan S, Bello Y. Dietary energy requirement of C juvenile at fixed crude protein and its effects on growth, nutrient performance, haematology and biochemical indices. Animal Research International. 2018;**15**(3):3090-3100

[121] Keremah RI, Beregha O. Effect of varying dietary protein levels on growth and nutrient utilization of African Catfish *Clarias gariepinus* fingerlings. Journal of Experimental Biology and Agricultural Sciences. JEBAS, India. 2014;**2**(1):13-18

[122] Craig SR, Helfrich LA, Kuhn D, Schwarz MH. Understanding Fish Nutrition, Feeds, and Feeding. Virginia Cooperative Extension, Virginia State University; 2017

[123] Hecht T. A review of on-farm feed management practices for North African catfish (*Clarias gariepinus*) in sub-Saharan Africa. In: On-farm Feeding and Feed Management in Aquaculture. FAO

Fisheries and Aquaculture Technical Paper No. 583. Rome: FAO; 2013. pp. 463-479

[124] Taj S, Han Q, Wu X, Yin H, Tian L, Yang H, et al. Effects of dietary protein-to-energy ratios on growth, immune response, antioxidative capacity, liver and intestinal histology, and growth-related gene expression in hybrid yellow catfish (*Pelteobagrus fulvidraco*♀× *Pelteobagrus vachelli*♂). Aquaculture Nutrition. 2023;**2023**(1):9106332

[125] Miles RD, Chapman FA. The concept of ideal protein in formulation of aquaculture feeds. In: Using Ideal Proteins in Aquaculture Feeds: Feeding Excess Protein Is Not Acceptable Lysine: The Reference Amino Acid. Gainesville, Florida (USA): IFAS Extension University of Florida; 2014. pp. 1-3

[126] Nhu B, Van Truong D, Thi TT, Duy TQ, That CT, Serrano AE. Lysine requirement of the spotted scat *Scatophagus argus* (Linaeus, 1766). Israeli Journal of Aquaculture-Bamidgeh. 2021;**73**:1-9

[127] Robinson E, Li MH, Hogue CD. Catfish nutrition: Nutrient requirements. Mississippi State University Extension Service Publication 2412; 2006

[128] FAO. North African catfish – Nutritional requirements. 2023. Available from: https://www.fao.org/fishery/affris/species-profiles/north-african-catfish/nutritional-requirements/en/

[129] Olugbenga O, Falaye AE, Kareem OK. Effect of phytase supplementation on the growth, mineral composition and phosphorus digestibility of African Catfish (*Clarias gariepinus*) juveniles. Animal Research International. 2017;**14**(2):2741-2750

[130] Mahboub HH, Shahin K, Zaglool AW, Roushdy EM, Ahmed SA. Efficacy of nano zinc oxide dietary supplements on growth performance, immunomodulation and disease resistance of African catfish *Clarias gariepinus*. Diseases of Aquatic Organisms. 2020;**142**:147-160

[131] Orire AM, Sadiku S. Effects of carbohydrate sources on the growth and body compositions of African Catfish (*Clarias gariepinus*) fingerlings. International Journal of Fisheries and Aquaculture. Academic Journals (Nigeria). 2014;**6**(5):55-61

[132] Shekarabi HSP, Shamsaie Mehrgan M, Banavreh A. Feasibility of superworm, *Zophobas morio*, meal as a partial fishmeal replacer in fingerling rainbow trout, *Oncorhynchus mykiss*, diet: Growth performance, amino acid profile, proteolytic enzymes activity and pigmentation. Aquaculture Nutrition. 2021;**27**(4):1077-1088

[133] Tippayadara N, Dawood MA, Krutmuang P, Hoseinifar SH, Doan HV, Paolucci M. Replacement of fish meal by black soldier fly (*Hermetia illucens*) larvae meal: Effects on growth, haematology, and skin mucus immunity of Nile tilapia, *Oreochromis niloticus*. Animals. 2021;**11**(1):193

[134] Hoffmann L, Rawski M, Pruszyńska-Oszmałek E, Kołodziejski P, Mazurkiewicz J. Environmentally sustainable feeding system for sea trout (*Salmotrutta M. trutta*): Live food and insect meal-based diets in larval rearing. Aquaculture Reports. 2021;**21**:100795

[135] Chemello G, Renna M, Caimi C, Guerreiro I, Oliva-Teles A, Enes P, et al. Partially defatted *Tenebrio molitor* larva meal in diets for grow-out rainbow trout, *Oncorhynchus mykiss* (Walbaum): Effects on growth performance, diet digestibility and metabolic responses. Animals. 2020;**10**(2):229

[136] Bruni L, Pastorelli R, Viti C, Gasco L, Parisi G. Characterisation of the intestinal microbial communities of rainbow trout (*Oncorhynchus mykiss*) fed with *Hermetia illucens* (black soldier fly) partially defatted larva meal as partial dietary protein source. Aquaculture. 2018;**487**:56-63

[137] Kroeckel S, Harjes AG, Roth I, Katz H, Wuertz S, Susenbeth A, et al. When a turbot catches a Fly: Evaluation of a pre-pupae meal of the black soldier fly (*Hermetia illucens*) as fish meal substitute—Growth performance and chitin degradation in juvenile turbot (*Psetta maxima*). Aquaculture. 2012;**364**:345-352

[138] Janssen RH, Vincken J, Van den Broek LA, Fogliano V, Lakemond CM. Nitrogen-to-protein conversion factors for three edible insects: *Tenebrio molitor*, *Alphitobius diaperinus*, and *Hermetia illucens*. Journal of Agricultural and Food Chemistry. 2017;**65**(11):2275-2278. DOI: 10.1021/acs.jafc.7b00471

[139] Pérez-Pacheco R, Hinojosa-Garro D, Ruíz-Ortíz F, Camacho-Chab JC, Ortega-Morales BO, Alonso-Hernández N, et al. Growth of the black soldier fly *Hermetia illucens* (Diptera: Stratiomyidae) on organic-waste residues and its application as supplementary diet for Nile tilapia *Oreochromis niloticus* (Perciformes: Cichlidae). Insects. 2022;**13**(4):326. DOI: 10.3390/insects13040326

[140] Ewald N, Vidakovic A, Langeland M, Kiessling A, Sampels S, Lalander C. Fatty acid composition of black soldier fly larvae (*Hermetia illucens*)–possibilities and limitations for modification through diet. Waste Management. 2020;**102**:40-47

[141] Gasco L, Oddon SB, Vandenberg GW, Veldkamp T, Biasato I. Factors affecting the decision-making process of using insect-based products in animal feed formulations. Journal of Insects as Food and Feed. 2023;**10**(10):1707-1718

[142] Fortes da Silva R, Martínez FJ, Villarroel M, Sanchez-Vazquez FJ. Daily feeding patterns and self-selection of dietary oil in Nile tilapia. Aquaculture Research. 2010;**42**(1):157-160

[143] Basto A, Matos E, Valente LM. Nutritional value of different insect larvae meals as protein sources for European sea bass (*Dicentrarchus labrax*) juveniles. Aquaculture. 2020;**521**:735085

[144] Panini RL, Freitas LEL, Guimarães AM, Rios C, da Silva MFO, Vieira FN, et al. Potential use of mealworms as an alternative protein source for Pacific white shrimp: Digestibility and performance. Aquaculture. 2017;**473**:115-120

[145] Marono S, Piccolo G, Loponte R, Di Meo C, Attia YA, Nizza A, et al. In vitro crude protein digestibility of *Tenebrio molitor* and *Hermetia illucens* insect meals and its correlation with chemical composition traits. Italian Journal of Animal Science. 2015;**14**(3):3889

[146] Fawole FJ, Adeoye AA, Tiamiyu LO, Ajala KI, Obadara SO, Ganiyu IO. Substituting fishmeal with *Hermetia illucens* in the diets of African catfish (*Clarias gariepinus*): Effects on growth, nutrient utilization, haemato-physiological response, and oxidative stress biomarker. Aquaculture. 2020;**518**:734849

[147] Hashizume A, Ido A, Ohta T, Thiaw ST, Morita R, Nishikawa M, et al. Housefly (Musca domestica) larvae preparations after removing the hydrophobic fraction are effective alternatives to fish meal in aquaculture

feed for red seabream (Pagrus major). Fishes. 2019;**4**(3):38

[148] Taufek NM, Aspani F, Muin H, Raji AA, Razak SA, Alias Z. The effect of dietary cricket meal (*Gryllus bimaculatus*) on growth performance, antioxidant enzyme activities, and haematological response of African catfish (*Clarias gariepinus*). Fish Physiology and Biochemistry. 2016;**42**:1143-1155

[149] Rema P, Saravanan S, Armenjon B, Motte C, Dias J. Graded incorporation of defatted yellow mealworm (*Tenebrio molitor*) in rainbow trout (*Oncorhynchus mykiss*) diet improves growth performance and nutrient retention. Animals. 2019;**9**(4):187

[150] Iaconisi V, Bonelli A, Pupino R, Gai F, Parisi G. Mealworm as dietary protein source for rainbow trout: Body and fillet quality traits. Aquaculture. 2017;**484**:197-204

[151] Coutinho F, Castro C, Guerreiro I, Rangel F, Couto A, Serra CR, et al. Mealworm larvae meal in diets for meagre juveniles: Growth, nutrient digestibility and digestive enzymes activity. Aquaculture. 2021;**535**:736362

[152] Jeong SM, Khosravi S, Yoon KY, Kim KW, Lee BJ, Hur SW, et al. Mealworm, *Tenebrio molitor*, as a feed ingredient for juvenile olive flounder, *Paralichthys olivaceus*. Aquaculture Reports. 2021;**20**:100747

[153] Rapatsa MM, Moyo NA. Evaluation of *Imbrasia belina* meal as a fishmeal substitute in *Oreochromis mossambicus* diets: Growth performance, histological analysis and enzyme activity. Aquaculture Reports. 2017;**5**:18-26

[154] Roncarati A, Cappuccinelli R, Meligrana MC, Anedda R, Uzzau S, Melotti P. Growing trial of gilthead sea bream (*Sparus aurata*) juveniles fed on chironomid meal as a partial substitution for fish meal. Animals. 2019;**9**(4):144

[155] Wang G, Peng K, Hu J, Yi C, Chen X, Wu H, et al. Evaluation of defatted black soldier fly (*Hermetia illucens* L.) larvae meal as an alternative protein ingredient for juvenile Japanese seabass (*Lateolabrax japonicus*) diets. Aquaculture. 2019;**507**:144-154

[156] Motte C, Rios A, Lefebvre T, Do H, Henry M, Jintasataporn O. Replacing fish meal with defatted insect meal (yellow mealworm *Tenebrio molitor*) improves the growth and immunity of Pacific white shrimp (*Litopenaeus vannamei*). Animals. 2019;**9**(5):258. DOI: 10.3390/ani9050258

[157] Rumbos CI, Mente E, Karapanagiotidis I, Vlontzos G, Athanassiou C. Insect-evaluation of different culture enclosures for musca domestica larval production and their utilization in Greece. Insects. 2021;**12**(2021):586

[158] Baldi L, Mancuso T, Peri M, Gasco L, Trentinaglia M. Consumer attitude and acceptance toward fish fed with insects: A focus on the new generations. Journal of Insects as Food and Feed. 2022;**8**(11):1249-1263. DOI: 10.3920/jiff2021.0109

[159] Sogari G, Amato M, Biasato I, Chiesa S, Gasco L. The potential role of insects as feed: A multi-perspective review. Animals. 2019;**9**(4):119

[160] Verbeke W. Profiling consumers who are ready to adopt insects as a meat substitute in a Western society. Food Quality and Preference. 2015;**39**:147-155

Chapter 4

Seaweed as a Support for the Development of Integrated Mariculture

Mostafa A.M. Soliman

Abstract

There is a growing tendency to broaden the scope of marine farming, with an ever-increasing focus on seafood as a crucial sector for supplying food to consumers. This has led to increased investment in this field by increasing the number of fish or cages used in seawater aquaculture. Which may affect the nature of the water in one way or another as a result of the increase in waste resulting from this industry? Therefore, the process of integration with seaweed, which is considered a natural biological filter, will help this industry flourish significantly. In addition to the potential for use as a nutritious food source for fish and humans.

Keywords: seaweeds, integration, mariculture, seafood, sustainability

1. Introduction

Human civilizations encounter the enormous challenge of providing food and livelihoods to a population of well over 9 billion people by the middle of the twenty-first century while also dealing with the disproportionate effects of climate change and resource degradation [1]. Approximately 70% of the surface of the Earth is covered by an uninterrupted line of seawater with an average depth of 5 kilometers. The oceans sustain floating forests with diverse sea plants and animals, and marine vegetation is thought to be more basic and richer than land-based plants [2]. Sustainable development is one of the nation's priorities. Sustainability is a concept used in industrial operations, especially mariculture, and it is part of the Sustainable Development Goals (SDGs) [3]. Mariculture is expected to provide new opportunities for the islands and coastal states' long-term economic growth [4]. Mariculture will be critical in satisfying these rising demands on a long-term basis [5].

Mariculture, commonly referred to as environmentally friendly aquaculture, is a type of aquatic culture practiced on salt water [6, 7]. The natural environment, socioeconomic status, organizational, and cultural factors are currently endangering the viability of mariculture [8–10]. Mariculture activities involve natural and human resources that produce outputs and services; their success must be financially lucrative for mariculture producers while also having benign social and environmental impacts [11]. The expansion of marine farming and the increase in fish stocking have impacted the environment surrounding the marine cages due to the intensive use of manufactured feed in feeding the fish. Therefore, it was necessary to turn to seaweed farming to eliminate the waste from the cages. Seaweeds are significant marine resources that contribute considerably to the sea's abundant biodiversity [2]. Seaweeds are macroalgae that live mostly in maritime environments. They are multicellular structures that can be seen with the naked eye and differ from higher plants in that they lack true roots, stalks, and leaves, making them thalloid in nature [2]. Cultivation of seaweed will be essential to this development since it will supply chemical extracts, animal feed, and human food sources. In addition, seaweed cultivation is a carbon-extraction activity, which increases its importance as a sustainable development activity. Currently, global seaweed output exceeds 30 million tonnes fresh weight [5], and production continues to rise. In addition, seaweed is preferable to the other approaches for a variety of reasons: (1) they may not only convert surplus nutrients (especially C, N, and P) into new biomass, but also improve DO, decrease CO_2, and enhance water pH in one move [12–14]; (2) unlike bivalve seaweed, which releases inorganic nutrients; (3) macroalgae are easier to work with than microalgae, microalgae populations are difficult to control because of bloom and crash cycles, and (4) seaweed also have higher productivity potentials that could be obtained if chemo-physical parameters (i.e., nutrients concentrations and flux, turbulence, irradiance, and temperature) could be properly managed [15–17]. Seaweeds are relatively analogous to aquaculture commodities, with unit values similar to those of aquatic animals and their products, which have very high trading values. Over the last 50 years, seaweeds have become the second-largest aquaculture product in terms of quantity, thanks to planned cultivation [1]. The aim of this chapter is to illustrate the effective role of seaweeds in supporting sustainability in several different forms and to encourage interest in expanding their production.

2. Role as biological filter

Mariculture continues to confront a number of challenges, including those related to its effects on the environment, the production of fodder, the kinds of species that are cultivated, and the selection of appropriate locations [18, 19]. Offshore aquaculture is becoming increasingly popular in coastal areas under heavy anthropogenic pressure as a means of mitigating the effects of eutrophication and overcoming space constraints [20–22]. But inshore mariculture production occurs all over the world in shallow areas with low hydrodynamic energy and near infrastructure on the mainland [23]; therefore, these plants have a significant impact on the surrounding ecosystem [24, 25]. Detritus from mariculture operations frequently has a detrimental impact on the environment, impacting water quality and biological communities [26]. Furthermore, according to Stabili et al. [27], uneaten feed causes disease outbreaks and deteriorates water quality, which adds to the growing worry over the prevalence of illnesses in the aquaculture sector globally. Because fish are given larger concentrations of antibiotics, aquaculture has frequently been criticized for being unsustainable and unfriendly to the environment [28, 29]. Over the past 20 years, research on aquaculture bioremediation has accelerated, and the moment is now right to advance associated technologies and lessen the environmental impact of this commercial activity. Integrated Multitrophic Aquaculture (IMTA) may be one of the most effective ways to counteract the effects of mariculture on inshore vegetation [30]. The increased nutrient load in the water column and sediments can be bioremediated in this system by the polyculture of several organisms from different trophic levels [31–33]. Seaweeds are essential for the development of submerged vegetation habitats in deep-sea, coastal, and estuarine environments, where light and nutrients are the only constraints on the growth and establishment of these communities [34–37]. Additionally, seaweeds are important to the ecology of aquatic habitats because they provide oxygen, serve as a nursery habitat for a range of marine animals, and provide food for a number of herbivores [36, 38]. They also support herbivorous species of animals (invertebrates like sea urchins and/or gastropods, and vertebrates like herbivorous fish) and give protection from carnivorous predators. They play an important role in the aquatic complex's trophic web [39, 40].

3. Role as food

Seaweeds have considerable potential and are now being used more and more globally. It is predicted that by 2100, there will be about 11 billion people on the planet, and more food production will be required to feed this expanding population. Seaweeds may provide a substitute food source for people and animals. Therefore, seaweed farming and wild harvesting, as well as creating novel culinary products derived from seaweed, are both feasible options for attaining sustainable development and economic growth in emerging economies [41]. Seaweeds that are edible are eaten raw, such as when added to salads, soups, or sushi wraps. According to Regulation (EU) 2015/2283, they can therefore be categorized as "novel foods" based on their nutritional makeup [42]. Asian nations have traditionally utilized algae as sea vegetables, particularly Japan, where an estimated 4–8 g of seaweed is ingested daily per person (dry weight). The three primary species that are consumed are wakame (*U. pinnatifida*), nori (*Porphyra tenera* and *Porphyra yezoensis*), and kombu (*L. japonica*) [43]. Seaweeds typically contain micronutrients like vitamins and minerals as well as macronutrients like lipids (important fatty acids, n-3 and n-6), proteins (essential amino acids), and carbs (dietary fiber). These ingredients have demonstrated a number of health advantages, including antibacterial, anti-inflammatory, anti-obesity, anticancer, and antioxidant properties [44]. Seaweed species differ in their chemical and biochemical composition [43]. According to Fleurence [45], this is particularly true for the protein content, which is high for some red seaweeds (up to 47% of dry weight) and low for numerous brown seaweeds (<15% of dry weight). Some species' protein content varies greatly according to the season. For example, the species *P. palmata*, sometimes known as dulse, has a protein content of 8% of dry weight at the end of summer and 25% at the start of spring [46].

3.1 The dietary composition of palatable seaweeds

Three brown and two red edible Spanish seaweeds' approximate content (moisture, ash, protein, and oil content), total dietary fiber content and physical and chemical properties are as follows: *Gigartina pistillata*, *Mastocarpus stellatus*, *Laminaria saccharina* (sweet kombu), *Bifurcaria bifurcata*, and *Himanthalia elongata* (sea spaghetti) were all examined. All samples had a high percentage of ashes (24.9–36.4%). *Laminaria* had significantly greater protein content (25.7%) than the red seaweeds (15.5–21.3%), with a range of 10.9–25.7%. Lipids (0.3–0.9%) were minor components in every sample, with the exception of Bifurcaria (5.6%). Finally, according to Gómez-Ordóñez et al. [47], these seaweeds are a rich source of food fiber, protein, and minerals for human consumption. In general, all seaweed species have extremely low-fat content, ranging from 0.9–4% of dry weight (**Table 1**) [48]. In contrast to protein content, lipid percentage displays no seasonal change, as has frequently been demonstrated for the edible seaweed *Grateloupia turuturu* [49, 50]. Algae are hence minimal in calories. Nonetheless, they typically have high carbohydrate content (**Table 1**). For some brown seaweed, such as *Eisenia bicyclis*, it can make up as much as 75% of their dry weight [51]. Algal carbohydrates are fiber-like, with fiber levels higher than those found in vegetables or fruits [52, 53]. Adding seaweeds increased ω-3 polyunsaturated fatty acids (PUFA) and decreased ω-6/ω-3 PUFA ratio. Nori increased the levels of serine, glycine, alanine, valine, tyrosine, phenylalanine, and arginine, whereas Wakame and Sea Spaghetti had no significant effect on amino acid profiles in the model systems [54]. A number of red (*Chondrus crispus* and *Porphyra tenera*) and brown (*Fucus vesiculosus, Laminaria digitata,* and *Undaria pinnatifida*) edible marine sea vegetables were tested for mineral content. High amounts of sulfate (1.3–5.9%) and ash (21.1–39.3%) were found in seaweeds. Compared to red algae (20.6–21.1%), brown algae had a greater ash concentration (30.1–39.3%). To help fulfill the daily requirements for certain vital minerals and trace elements, edible brown and red seaweeds could be utilized as a dietary supplement [55].

It is not surprising that modern consumers prefer fresh, lightly preserved foods free of chemical preservatives but containing natural compounds that may improve their health because they are well aware of the nutritional value of food and the negative effects that synthetic preservatives may have on their health [56]. Algal constituents, including polysaccharides, phenolic compounds, bioactive peptides, pigments, omega-3 fatty acids, terpenoids, and vitamins, have recently gained attention for their powerful biological effects [57, 58]. Aside from antihypertensive, antioxidant, antithrombotic, antimicrobial, and immunomodulatory effects [59], these components demonstrated a wide range of potential applications in the food, therapeutic, and drug industries, as well as in the development of functional ingredients. Seaweeds include natural antioxidants and antibacterial agents that could be used as nutraceuticals, thereby improving food quality and safety.

Content (% Dry Weight)	***Undaria pinnatifida*** **(wakame)**	***Laminaria digitata*** **(Kombu Breton)**	***Porphyra tenera*** **(Nori)**	***Gracilaria verrucosa*** **(Ogonori)**	***Ulva lactuca*** **(sea lettuce)**
Water	10–18	13–16	5–11	12–14	15–22
Carbohydrate	38–48 t	42–62	44–56	55–61	41–60
Total fiber	35.3	37.3	34.7	—	38.1
Soluble fiber	30.3	32.6	17.9	—	21.3
Insoluble fiber	5.3	4.7	16.8	—	16.8
Protein	11–14	5–9	29–36	7–9	10–23
Lipid	0.5–2	0.9–2.1	0.6–0.7	0.1–1	1–3.5
Ash	26–32	19–36	8–10	15–20	14–29
Calcium	1.1–3	0.5–3	—	—	0.8–5.6
Magnesium	1–3	0.5–2	—	—	2–5.2

Table 1.
Some edible seaweeds' chemical and biochemical composition.

4. Role as diet ingredient

Marine macroalgae (seaweeds) are a nutritious feed supplement that provides essential amino acids, fatty acids, beneficial polysaccharides, antioxidants, minerals, and vitamins [60, 61]. Herbivorous fishes prefer them as food due to their low pH and specialized intestines for plant digestion [62]. Furthermore, they enhance the immune system, have antiviral and antibacterial properties, improve gut function, and are stress resistant, making them a viable alternative to fish meal and helping to relieve pressure on wild fish populations [63]. Because macroalgal polysaccharides can alter digestibility, they directly affect how well nutrients are assimilated in fish guts, making them essential to the feeding process [64]. *Salmo salar* lysozyme activity was stimulated by alginate derived from *Ascophyllum nodosum* [65]. In addition to its nutritional value, seaweed contains bioactive compounds that have antimicrobial, antiviral, antioxidative, anti-inflammatory, and neuroprotective properties. These chemicals serve as scavengers of reactive oxygen species (ROS) and enhance immune response and stress resistance [66]. *Sargassum wightii* fucoidan boosted immunological markers like *Pangasianodon hypophthalmus* respiratory burst activity, total leucocyte count, and phagocytic activity [67].

The use of seaweeds as an element in aquafeed for various fish species has been the subject of numerous studies recently. For instance, but not exclusively. The effects of red algae (*Pterocladia capillacea*) and green algae (*Ulva lactuca*) at three different levels (0.0, 2.5, and 5%) on the blood indices, growth rates, feed efficiency, and carcass composition of *Oreochromis niloticus* fingerlings were investigated by Khalafalla and El-Hais [68]. The findings demonstrated that serum total protein, globulin, albumin, and liver enzyme activity did not significantly change. Nile tilapia fed diets containing 2.5 and 5% red and green algae supplementation had considerably increased growth rate and feed efficiency values. When compared to other diets, fish fed a diet containing 5% green algae (*Ulva lactuca*) grew at respectable rates. The effects of supplemented diets for Nile tilapia were negligible, with very little increases and losses in carcass lipids and protein. Moreover, Garcia-Casal et al. [69] found that 5% *U. rigida* food supplementation improved Nile tilapia development and nutritional use.

Furthermore, Guroy et al. [69] discovered that *Oreochromis niloticus* fed diets enriched with varying quantities of *Ulva* meal gained more weight (5 to 10%). Incorporating 5% green seaweed (*Ulva lactuca*) in Nile tilapia (*Oreochromis niloticus*), diets improved growth, food utilization, and immunological response [70]. Saleh [71] investigated the impact of fresh macroalgae or seaweeds (*Enteromorpha flaxuse*) on the development, survival, and feed efficiency of young hybrid red tilapia. Red tilapia juveniles were fed one of three diets: artificial feed only, fresh macroalgae only, or 50% artificial feed and 50% fresh macroalgae. The fish with the highest final weight and specific growth rate were fed only artificial feed or fresh algae. However, fish fed only artificial feed grew longer, gained more weight, and had a faster daily growth rate. Red tilapia fed exclusively with artificial feed had the highest feed conversion ratio. Red tilapia juveniles were not as accepting of feeding on fresh algae as they were of feeding on artificial feed, which may explain the reduced growth in this treatment. The survival rate ranged from 86 to 90%, with no notable differences between therapies. Additionally, Costa et al. [72] found that both fresh and dried brown seaweed can be fed to fish juveniles like gigantic gourami (*Osphronemus goramy*), spotted scat (*Scatophagus argus*), and red tilapia (*Oreochromis* sp.) in place of commercial diets. The growth performance of Indian major carp (Catla catla) over time was examined in a study by Kotnala et al. [73] using feeds made with three seaweeds: *Padina tetrastomatica*, *Chlorodesmis fastigiata*, and *Stoechospermum marginatum*. The findings showed that seaweeds such as *P. tetrastomatica* and *C. fastigiata* could be added to commercially prepared diets to improve the growth of large carp fingerlings.

5. Role in medicine

Although seaweeds include a variety of elements, some chemicals are specific to particular algal classes or species, or the combination of certain bioactive compounds is uncommon and improves the overall effect. Season, harvesting time, location, abiotic (salinity, pH, temperature, water nutrient composition), and biotic (herbivory or direct competition with other benthic organisms) factors all affect the amount and quality of biologically active compounds in seaweeds [74, 75]. Due to the enormous changes in the environment brought about by global climate change, seaweed habitats are characterized by a variety of intricate and dynamic characteristics. The concentration and caliber of the chemical compounds generated are significantly impacted by each of these variables. Thus, each of these has a significant impact on the concentration and composition of chemical compounds generated, or just the quality [76, 77].

5.1 Properties of antioxidants

Based on their antioxidant mechanisms, primary and secondary antioxidants have been categorized. Lipid oxidation is largely prevented by the major antioxidants' direct reactivity with free radicals, which changes them into more stable, non-radical compounds. Through processes like oxygen scavenging, singlet-oxygen quenching (in photooxidation), and transition metal chelation, secondary antioxidants stop or restrict lipid oxidation [78, 79]. The phytochemical composition of seaweed extracts has been thoroughly investigated as a natural antioxidant source in recent years. The high antioxidant activity of phlorotannins, a unique class of chemicals, has been discovered [79, 80, 81]. Phlorotannins may scavenge free radicals, including superoxide, peroxide, and nitric radicals, as well as chelate ferrous ions [80, 82]. Phlorotannin structures are extremely complex, making characterization and identification difficult and infrequently reported [79].

5.2 Properties of antimicrobial

The food industry has been searching for new natural antimicrobials in recent years to produce safe foods and extend their shelf life. To be approved for industrial manufacturing, these natural substances must, however, function on par with or better than conventional, synthetic preservatives while maintaining or enhancing the product's sensory qualities [83]. Additionally, customers must be informed about the new antimicrobials used in the new food products [84]. Research indicates that seaweed extracts have antibacterial properties against foodborne pathogens and spoilage bacteria [85]. The use of extracts or seaweed powder in food models, however, was only mentioned in a small number of researches. These investigations often use changes in total viable count (TVC) to examine the antimicrobial effect and microbiological quality, as well as whether the entire microbial community can flourish at room temperature with oxygen present. Factors to be taken into account for the next studies that will allow seaweeds to be used in the sector. Challenge tests on foodborne pathogens are conducted to determine their significance in food safety, along with safety concerns. Because of heavy metals, pesticide residues, dioxin, pharmaceuticals, or marine biotoxins, the use of whole seaweed powder in food products without extraction or purification should be reexamined [83, 85, 86]. Chemical identification and purification of used crude extracts, which are necessary for their industrial applications, are absent from the majority of studies. The antibacterial activity of extracts is undeniable in spite of this. Seaweeds' antibacterial activity is attributed to phenoltannins, polysaccharides, pigments, proteins, fatty acids, halogenated chemicals, and other trace amounts of other substances [85]. The antimicrobial mechanisms of action of various categories of chemicals vary, but the majority are still unknown [83]. It can be challenging and complex to identify antimicrobial compounds in extracts, and it's important to take into account the combined effects of isolated substances [87]. To create extracts with the maximum antibacterial activity, care should also be taken to select the optimum drying and extraction techniques with the ideal parameters for each seaweed [88].

6. Conclusions

Interest in seaweed cultivation is not only a commercial or profitable trend but also an environmentally friendly trend. In addition, it is one of the main pillars supporting the production of farmed seafood. In addition to preserving the marine environment without affecting the biodiversity of the sea, it is necessary to strongly move toward its cultivation in the seas and oceans in an intensive and organized manner.

Author details

Mostafa A.M. Soliman
Animal System Production Research Department, Animal Production Research Institute (APRI), Agriculture Research Center, Ministry of Agriculture, Dokky-Giza-Egypt

*Address all correspondence to: mostafa.soliman242@gmail.com

References

[1] Food and Agriculture Organization of the United Nations. The state of world fisheries and aquaculture 2018. In: Meeting the Sustainable Development Goals. Rome: Food and Agriculture Organization of the United Nations; 2018. 227 pp

[2] Baweja P, Kumar S, Sahoo D, Levine I. Biology of seaweeds. In: Seaweed in Health and Disease Prevention. Chapter (3). 2016. pp. 41-106

[3] Eriksson H, Troell M, Brugere C, Chadag V, Phillips M, Andrew NL. Equitable Mariculture: A Diagnostic Framework for Equitable Mariculture Development in the Western Indian Ocean. Australian Centre for International Agricultural Research (ACIAR); Report, WorldFish Headquarters (Malaysia). 2018

[4] Islam MM, Shamsuddoha MD. Coastal and marine conservation strategy for Bangladesh in the context of achieving blue growth and sustainable development goals (SDGs). Environmental Science & Policy. 2018;**87**:45-54

[5] Food and Agriculture Organization of the United Nations. Fisheries and aquaculture statistics. 2017. Available from: http://www.fao.org/fishery/en [Accessed: August 30, 2019]

[6] Ministry of Marine Affairs and Fisheries. Marine and Fisheries Information Statistics. Pusat Data: Statistik dan Informasi, Jakarta, Kementerian Kelautan dan Perikanan; 2020. 590 p

[7] Pomeroy RS, Parks JE, Balboa CM. Farming the reef: Is aquaculture a solution for reducing fishing pressure on coral reefs? Marine Policy. 2006;**30**(2):111-130

[8] Fredheim A, Reve T. Future prospects of marine aquaculture. Oceans 2018 MTS/IEEE Charleston. Institute of Electrical and Electronics Engineers (IEEE); 2018. pp. 1-8

[9] Marzuki M, Nurjaya IW, Purbayanto A, Budiharso S, Supriyono E. Sustainability analysis of Mariculture management in Saleh Bay of Sumbawa District. Environmental Management and Sustainable Development. 2014;**3**(2):127

[10] Yu J, Yin W. Exploring stakeholder engagement in mariculture development: Challenges and prospects for China. Marine Policy. 2019;**103**:84-90

[11] Liu Y, Sumaila UR. Economic analysis of netcage versus sea-bag production systems for salmon aquaculture in British Columbia. Aquaculture Economics & Management. 2007;**11**(4):371-395

[12] Hirata H, Yamasaki S, Maenosono H, Nakazono T, Yamauchi T, Matsuda M. Relative budgets of p O_2 and p CO_2 in cage polycultured Red Sea bream, Pagrus major and sterile Ulva sp. Aquaculture Science. 1994;**42**(2):377-381

[13] Rai LC, Kumar HD, Mohn FH, Soeder CJ. Services of algae to the environment. Journal of Microbiology and Biotechnology. 2000;**10**(2):119-136

[14] Chopin T, Buschmann AH, Halling C, Troell M, Kautsky N, Neori A, et al. Integrating seaweeds into marine aquaculture systems: A key toward sustainability. Journal of Phycology. 2001;**37**(6):975-986

[15] Goldman JC. Outdoor algal mass cultures—II. Photosynthetic yield limitations. Water Research. 1979;**13**(2): 119-136

[16] Wetzel RG. Limnology. Fort Worth, USA: Saunders College Publishing; 1983. p. 767

[17] Pagand P, Blancheton JP, Lemoalle J, Casellas C. The use of high rate algal ponds for the treatment of marine effluent from a recirculating fish rearing system. Aquaculture Research. 2000;**31**(10):729-736

[18] Serpa D, Duarte P. Impacts of aquaculture and mitigation measures. Dynamic Biochemistry, Process Biotechnology and Molecular Biology. 2008;**2**:1-20

[19] Costello C, Cao L, Gelcich S, Cisneros-Mata MÁ, Free CM, Froehlich HE, et al. The future of food from the sea. Nature. 2020;**588**(7836):95-100

[20] Ministerio de Medio Ambiente y Medio Rural y Marino. IUCNa-the World Conservation Union, Federation of European Aquaculture Producers. Spain: Aquaculture Site Selection and Site Management. IUCN; 2009

[21] Ministerio de Medio Ambiente y Medio Rural y Marino. IUCNb-the World Conservation Union, Federation of European Aquaculture Producers. Spain: Aquaculture: Responsible Practices and Certification. IUCN; 2009

[22] Mizuta DD, Dixon MS, Maney EJ Jr, Fregeau M, Wikfors GH. Offshore mussel aquaculture: Strategies for farming in the changing environment of the northeast US shelf EEZ. Bulletin of Japan Fisheries Research and Education Agency. 2019;**49**:111-119

[23] Osmundsen TC, Amundsen VS, Alexander KA, Asche F, Bailey J, Finstad B, et al. The operationalisation of sustainability: Sustainable aquaculture production as defined by certification schemes. Global Environmental Change. 2020;**60**:102025

[24] Muir JF, Brugere C, Young JA, Stewart JA. The solution to pollution? The value and limitations of environmental economics in guiding aquaculture development. Aquaculture Economics & Management. 1999;**3**(1):43-57

[25] Gentry RR, Lester SE, Kappel CV, White C, Bell TW, Stevens J, et al. Offshore aquaculture: Spatial planning principles for sustainable development. Ecology and Evolution. 2017;7(2):733-743. DOI: 10.1002/ece3.2637

[26] Wang S, Yan Z, Wang P, Zheng X, Fan J. Comparative metagenomics reveals the microbial diversity and metabolic potentials in the sediments and surrounding seawaters of Qinhuangdao mariculture area. PLoS One. 2020;**15**(6):e0234128. DOI: 10.1371/journal.pone.0234128

[27] Stabili L, Schirosi R, Licciano M, Mola E, Giangrande A. Bioremediation of bacteria in aquaculture waste using the polychaete Sabella spallanzanii. New Biotechnology. 2010;**27**(6):774-781. DOI: 10.1016/j.nbt.2010. 06.018

[28] González-Gaya B, García-Bueno N, Buelow E, Marin A, Rico A. Effects of aquaculture waste feeds and antibiotics on marine benthic ecosystems in the Mediterranean Sea. Science of the Total Environment. 2022;**806**:151190. DOI: 10.1016/j.scitotenv.2021.151190

[29] Wang X, Lin Y, Zheng Y, Meng F. Antibiotics in mariculture systems: A review of occurrence, environmental behavior, and ecological effects. Environmental Pollution. 2022;**293**:118541. DOI: 10.1016/j.envpol.2021.118541

[30] Buck BH, Troell MF, Krause G, Angel DL, Grote B, Chopin T. State of the art and challenges for offshore integrated multi-trophic aquaculture (IMTA). Frontiers in Marine Science. 2018;**5**:165. DOI: 10.3389/fmars.2018.00165

[31] Fang J, Jiang Z, Jansen HM, Hu F, Fang J, Liu Y, et al. Applicability of Perinereis aibuhitensis Grube for fish waste removal from fish cages in Sanggou Bay, PR China. Journal of Ocean University of China. 2017;**16**:294-304. DOI: 10.1007/s11802-017-3256-1

[32] Li J, Zhang W, Ding J, Xue S, Huo E, Ma Z, et al. Effect of large-scale kelp and bivalve farming on seawater carbonate system variations in the semi-enclosed Sanggou Bay. Science of the Total Environment. 2021;**753**:142065. DOI: 10.1016/j.scitotenv.2020.142065

[33] Naskar S, Biswas G, Kumar P, De D, Sawant PB, Das S, et al. Effects of estuarine oyster, *Crassostrea cuttackensis* as the extractive species at varied densities on productivity and culture environment in brackishwater integrated multi-trophic aquaculture (BIMTA) system. Aquaculture. 2022;**554**:738128. DOI: 10.1016/j.aquaculture.2022.738128

[34] Heck KL Jr, Hays G, Orth RJ. Critical evaluation of the nursery role hypothesis for seagrass meadows. Marine Ecology Progress Series. 2003;**253**:123-136

[35] Lefcheck JS, Hughes BB, Johnson AJ, Pfirrmann BW, Rasher DB, Smyth AR, et al. Are coastal habitats important nurseries? A meta-analysis. Conservation Letters. 2019;**12**(4):e12645

[36] García-Poza S, Pacheco D, Cotas J, Marques JC, Pereira L, Gonçalves AM. Marine macroalgae as a feasible and complete resource to address and promote sustainable development goals (SDGs). Integrated Environmental Assessment and Management. 2022;**18**(5):1148-1161

[37] García-Poza S, Cotas J, Morais T, Pacheco D, Pereira L, Marques JC, et al. Global trade of seaweed foods. In: Sustainable Global Resources of Seaweeds: Food, Pharmaceutical and Health Applications. Vol. 2. Cham: Springer International Publishing; 2022. pp. 325-337

[38] Pfister CA, Altabet MA, Weigel BL. Kelp beds and their local effects on seawater chemistry, productivity, and microbial communities. Ecology. 2019;**100**(10):e02798

[39] Butt KR, Méline C, Pérès G. Marine macroalgae as food for earthworms: Growth and selection experiments across ecotypes. Environmental Science and Pollution Research. 2020;**27**(27):33493-33499

[40] Randall J, Johnson CR, Ross J, Hermand JP. Acoustic investigation of the primary production of an Australian temperate macroalgal (Ecklonia radiata) system. Journal of Experimental Marine Biology and Ecology. 2020;**524**:151309

[41] Gajaria TK, Mantri VA. Perspectives and attitudes towards the functional and safety aspects of seaweeds for edible applications in India. Food. 2021;**10**(12):3026

[42] European Parliament and the Council of the European Union. Regulation (EU) 2015/2283 of the European Parliament and of the council of 25 November 2015 on novel foods, amending regulation (EU) No 1169/2011 of the European Parliament and of the council and repealing regulation (EC) No 258/97 of the European Parliament and of the council and commission regulation (EC) No 1852/2001. Official Journal of the European Union. 2015;**327**:1-22

[43] Fleurence J. Seaweeds as food. Seaweed in Health and Disease Prevention. Cambridge, MA, USA: Academic Press; 2016;**1**:149-167

[44] Ms VQ, Miss MS, Miss GG, Ms PJ. Seaweeds in bakery and farinaceous foods: A mini-review. International Journal of Gastronomy and Food Science. 2022;**28**:100403

[45] Fleurence J. Seaweed proteins. Proteins in Food Processing. 2004;**197**:211

[46] Galland-Irmouli AV, Fleurence J, Lamghari R, Luçon M, Rouxel C, Barbaroux O, et al. Nutritional value of proteins from edible seaweed Palmaria palmata (Dulse). The Journal of Nutritional Biochemistry. 1999;**10**(6):353-359

[47] Gómez-Ordóñez E, Jiménez-Escrig A, Rupérez P. Dietary fibre and physicochemical properties of several edible seaweeds from the northwestern Spanish coast. Food Research International. 2010;**43**(9):2289-2294

[48] Dawczynski C, Schubert R, Jahreis G. Amino acids, fatty acids, and dietary fibre in edible seaweed products. Food Chemistry. 2007;**103**(3):891-899

[49] Denis C, Morançais M, Li M, Deniaud E, Gaudin P, Wielgosz-Collin G, et al. Study of the chemical composition of edible red macroalgae Grateloupia turuturu from Brittany (France). Food Chemistry. 2010;**119**(3):913-917

[50] Kendel M, Couzinet-Mossion A, Viau M, Fleurence J, Barnathan G, Wielgosz-Collin G. Seasonal composition of lipids, fatty acids, and sterols in the edible red alga Grateloupia turuturu. Journal of Applied Phycology. 2013;**25**:425-432

[51] Lahaye M. Marine algae as sources of fibres: Determination of soluble and insoluble dietary fibre contents in some 'sea vegetables'. Journal of the Science of Food and Agriculture. 1991;**54**(4):587-594

[52] Darcy-Vrillon B. Nutritional aspects of the developing use of marine macroalgae for the human food industry. International Journal of Food Sciences and Nutrition. 1993;**44**:23-35

[53] MacArtain P, Gill CI, Brooks M, Campbell R, Rowland IR. Nutritional value of edible seaweeds. Nutrition Reviews. 2007;**65**(12):535-543

[54] López-López I, Bastida S, Ruiz-Capillas C, Bravo L, Larrea MT, Sánchez-Muniz F, et al. Composition and antioxidant capacity of low-salt meat emulsion model systems containing edible seaweeds. Meat Science. 2009;**83**(3):492-498

[55] Rupérez P. Mineral content of edible marine seaweeds. Food Chemistry. 2002;**79**(1):23-26

[56] Petruzzi L, Corbo MR, Sinigaglia M, Bevilacqua A. Microbial spoilage of foods: Fundamentals. In: Bevilacqua A, Corbo MR, Sinigaglia M, editors. The Microbiological Quality of Food Foodborne Spoilers. Woodhead Publishing., MA: Elsevier; 2017. pp. 1-21. DOI: 10.1016/B978-0-08-100502-6.00002-9

[57] Gullón B, Gagaoua M, Barba FJ, Gullón P, Zhang W, Lorenzo JM. Seaweeds as promising resource of bioactive compounds: Overview of novel extraction strategies and design of tailored meat products. Trends in Food Science & Technology. 2020;**100**:1-8. DOI: 10.1016/j.tifs.2020.03.039

[58] Šimat V, Čagalj M, Skroza D, Gardini F, Tabanelli G, Montanari C,

et al. Sustainable sources for antioxidant and antimicrobial compounds used in meat and seafood products. In: Advances in Food and Nutrition Research. Vol. 97. 2021. pp. 55-118. DOI: 10.1016/bs.afnr.2021.03.001

[59] Lafarga T, Acién-Fernández FG, Garcia-Vaquero M. Bioactive peptides and carbohydrates from seaweed for food applications: Natural occurrence, isolation, purification, and identification. Algal Research. 2020;**48**:101909. DOI: 10.1016/j.algal.2020.101909

[60] Guedes AC, Sousa-Pinto I, Malcata FX. Application of microalgae protein to aquafeed. In: Handbook of Marine Microalgae, Biotechnology Advances. Academic Press; 2015. pp. 93-125

[61] Ismail MM, El Zokm GM, El-Sayed AA. Variation in biochemical constituents and master elements in common seaweeds from Alexandria coast, Egypt, with special reference to their antioxidant activity and potential food uses: Prospective equations. Environmental Monitoring and Assessment. 2017;**189**:1-7

[62] Horn MH, Messer KS. Fish guts as chemical reactors: A model of the alimentary canals of marine herbivorous fishes. Marine Biology. 1992;**113**:527-535

[63] Cyrus MD, Bolton JJ, Scholtz R, Macey BM. The advantages of Ulva (*Chlorophyta*) as an additive in sea urchin formulated feeds: Effects on palatability, consumption and digestibility. Aquaculture Nutrition. 2015;**21**(5):578-591

[64] Sinha AK, Sinha AK, Harinder-Makkar H, De Boeck G, Becker K. Nonstarch polysaccharide in human nutrition- a review. Critical Reviews in Food Science and Nutrition. 2012;**52**(10):899-935

[65] Van Doan H, Doolgindachbaporn S, Suksri A. Effects of low molecular weight agar and lactobacillus plantarum on growth performance, immunity, and disease resistance of basa fish (Pangasius bocourti, Sauvage 1880). Fish & Shellfish Immunology. 2014;**41**(2):340-345

[66] Prabu DL, Sahu NP, Pal AK, Dasgupta S, Narendra A. Immunomodulation and interferon gamma gene expression in sutchi cat fish, Pangasianodon hypophthalmus: Effect of dietary fucoidan rich seaweed extract (FRSE) on pre and post challenge period. Aquaculture Research. 2016;**47**(1):199-218

[67] Undeland I. Oxidative stability of seafood. In: Oxidative Stability and Shelf Life of Foods Containing Oils and Fats. AOCS Press; 2016. pp. 391-460

[68] Khalafalla MM, El-Hais AE. Evaluation of seaweeds Ulva rigida and Pterocladia capillaceaas dietary supplements in Nile tilapia fingerlings. Journal of Aquaculture Research & Development. 2015;**6**:312. DOI: 10.4172/2155-9546.1000312

[69] García-Casal MN, Pereira AC, Leets I, Ramírez J, Quiroga MF. High iron content and bioavailability in humans from four species of marine algae. The Journal of Nutrition. 2007;**137**(12):2691-2695

[70] Natify W, Droussi M, Berday A, Araba M. Effect of the seaweed Ulva lactuca as a feed additive on growth performance, feed utilization and body composition of Nile tilapia (*Oreochromis niloticus* L.). International Journal of Agronomy and Agricultural Research. 2015;7(3):85-92

[71] Saleh HS. Can artificial feed be replace by fresh macro algae (*Enteromorpha flaxusa*) in hybrid red tilapia (*Oreochromis mossambicus* × *Oreochromis niloticus*) juvenile nutrition. Journal of Oceanography and Marine Research. 2020;**8**:200

[72] Costa MM, Oliveira ST, Balen RE, Bueno Junior G, Baldan LT, Silva LC, et al. Brown seaweed meal to Nile tilapia fingerlings. Archivos de Zootecnia. 2013;**62**(237):101-109

[73] Kotnala S, Dhar P, Das P, Chatterji A. Growth performance of fingerlings of the Indian major carp Catla catla (Ham.) fed with feeds supplemented with different seaweeds. Pertanika Journal of Science and Technology. 2010;**18**(2):255-262

[74] Cotas J, Leandro A, Pacheco D, Gonçalves AM, Pereira L. A comprehensive review of the nutraceutical and therapeutic applications of red seaweeds (Rhodophyta). Life. 2020;**10**(3):19. DOI: 10.3390/life10030019

[75] Pacheco D, Araújo GS, Cotas J, Gaspar R, Neto JM, Pereira L. Invasive seaweeds in the Iberian Peninsula: A contribution for food supply. Marine Drugs. 2020;**18**(11):560. DOI: 10.3390/md18110560

[76] Leandro A, Pereira L, Gonçalves AM. Diverse applications of marine macroalgae. Marine Drugs. 2019;**18**(1):17. DOI: 10.3390/md18010017

[77] Gaubert J, Payri CE, Vieira C, Solanki H, Thomas OP. High metabolic variation for seaweeds in response to environmental changes: A case study of the brown algae Lobophora in coral reefs. Scientific Reports. 2019;**9**(1):993. DOI: 10.1038/s41598-018-38177-z

[78] Decker EA. Antioxidant mechanisms. In: Akoh CC, Min DB, editors. Food Lipids: Chemistry, Nutrition, and Biotechnology. New York: Marcel Dekker; 2002. pp. 397-422

[79] Hermund DB. Antioxidant properties of seaweed-derived substances. In: Bioactive Seaweeds for Food Applications. Natural Ingredients for Healthy Diets. Academic Press; 2018. pp. 201-221. DOI: 10.1016/B978-0-12-813312-5.00010-8

[80] Generalić Mekinić I, Skroza D, Šimat V, Hamed I, Čagalj M, Popović PZ. Phenolic content of brown algae (Pheophyceae) species: Extraction, identification, and quantification. Biomolecules. 2019;**9**(6):244. DOI: 10.3390/biom9060244

[81] Jacobsen C, Sørensen AD, Holdt SL, Akoh CC, Hermund DB. Source, extraction, characterization, and applications of novel antioxidants from seaweed. Annual Review of Food Science and Technology. 2019;**10**(1):541-568. DOI: 10.1146/annurev-food-032818-121401

[82] Roohinejad S, Koubaa M, Barba FJ, Saljoughian S, Amid M, Greiner R. Application of seaweeds to develop new food products with enhanced shelf life, quality and health-related beneficial properties. Food Research International. 2017;**99**:1066-1083. DOI: 10.1016/j.foodres.2016.08.016

[83] Cabral EM, Oliveira M, Mondala JR, Curtin J, Tiwari BK, Garcia-Vaquero M. Antimicrobials from seaweeds for food applications. Marine Drugs. 2021;**19**(4):211. DOI: 10.3390/md19040211

[84] Polat S, Trif M, Rusu A, Šimat V, Čagalj M, Alak G, et al. Recent advances in industrial applications of seaweeds. Critical Reviews in Food Science and Nutrition. 2023;**63**(21):4979-5008

[85] Pérez MJ, Falqué E, Domínguez H. Antimicrobial action of compounds from marine seaweed. Marine Drugs. 2016;**14**(3):52. DOI: 10.3390/md14030052

[86] Banach JL, Hoek-van den Hil EF, van der Fels-Klerx HJ. Food safety hazards in the European seaweed chain. Comprehensive Reviews in Food Science and Food Safety. 2020;**19**(2):332-364. DOI: 10.1111/1541-4337.12523

[87] Skroza D, Šimat V, Smole Možina S, Katalinić V, Boban N, Generalić MI. Interactions of resveratrol with other phenolics and activity against food-borne pathogens. Food Science & Nutrition. 2019;7(7):2312-2318. DOI: 10.1002/fsn3.1073

[88] Čagalj M, Skroza D, Tabanelli G, Özogul F, Šimat V. Maximizing the antioxidant capacity of Padina pavonica by choosing the right drying and extraction methods. PRO. 2021;**9**(4):587. DOI: 10.3390/pr9040587

Chapter 5

The Potential of the Inland Fisheries Sector in South Africa and Its Role in the Rural Economy

Esau Matthews Mbokane and Hlulani Archiebold Hlungwani

Abstract

The contributions of inland fisheries to food security, livelihood provision, poverty alleviation, and economic development in many developing countries is well-known. Inland fisheries can lead to the development of strategies to alleviate poverty and enhance food security. Efforts to grow the sector have increased in Africa after it was recognised as a priority investment area for regional economic development and poverty eradication. The South African government has recently recognised the sector as having the potential to help achieve the 2030 Sustainable Development Goals of the UN, including hunger and poverty. However, in South Africa, inland fisheries is largely informal in nature and its socioeconomic significance is frequently ignored in economics statistics. A number of studies have highlighted several factors as hindering the successful development of the sector. These include, among others, lack of policy, insufficient data on the inland fisheries potential of dams, and lack of training of rural communities on the economic potential of inland fisheries. Therefore, improving the contribution of inland fisheries to the socioeconomic well-being and lives of rural populations requires addressing these issues. This chapter discusses the current state of the inland fisheries sector in South Africa and identifies areas that need to be addressed urgently to improve the economic potential of the sector for the benefit of poor communities. The chapter provides an overview of the potential of the inland fisheries sector in South Africa and its role in growing the rural economy.

Keywords: economic growth, food security, poverty alleviation, rural development, policy development

1. Introduction

The inland fisheries sector is one of the most important industries with a significant potential in boosting the socio-economic status of rural communities in developing countries [1]. In Africa, inland fisheries are recognised as means of promoting rural development, poverty alleviation, food security, and improving livelihoods [1, 2]. Consequently, the African Union has designated this industry as a priority investment area [3]. In rural areas, where alternative employment opportunities may be limited, the sector offers a source of income, food, employment. In South Africa,

IntechOpen

the economic and social potential of inland fisheries is also gaining traction. The government regards the fisheries sector as a panacea to poverty and arresting rising levels of unemployment, especially poor communities. However, despite numerous attempts to develop the sector, beginning in the 1970s, the fisheries sector is still underdeveloped [4]. Concerns about the sustainability of available stocks and future economic viability have been highlighted as some of the factors hampering its full development. Importantly, South Africa lacked an inland fisheries policy until 2021. Other possible reasons include the lack of information on the history, present state, and potential or the sustainability of the inland resource, lack of access to fishing gear and marketing of fish products [4, 5]. Therefore, the government's water resource management plans did not take into account the potential socioeconomic value of fishing [5]. In this chapter, we highlight the sector's economic potential as well as several obstacles that may be preventing it from expanding in a sustainable manner in South Africa.

2. History of inland fisheries in South Africa

In the early 1960s, the South African government build several hatcheries around the country to stock inland waters with fish species. The aim was to conserve threatened species, promote aquaculture, but more importantly, to promote recreational and consumptive fisheries. However, the focus has been on stocking public waters with indigenous species to conserve their genetic diversity. Since the 1970s and early 1980s, efforts have been made to set up catch fisheries in inland waters. Such efforts started with stock enhancement using mullet in the Eastern Cape. Two mullet species, *Myxus capensis* and *Mugil cephalus*, were introduced into many Eastern Cape impoundments [6]. The fingerlings were placed into impoundments after being wild caught in Eastern Cape estuaries. It was believed then that mullets were more commercially viable than the *Labeo umbratus* and *Cyprinus carpio*. Therefore, the stocking was intended to improve the fisheries potential of impoundments in Eastern Cape for gillnet fisheries [6]. During this period, similar attempts were made to promote inland fisheries by stocking impoundments using hatchery-bred fingerlings in impoundments in the Eastern Cape, KwaZulu Natal, Free State, Limpopo and North-West [7–13]. To promote the stocking of the dams in these areas, the South African government build several hatcheries around the country [13, 14]. The aim was also to conserve threatened species, promote aquaculture, but more importantly, to promote recreational and consumptive fisheries.

3. Current status of inland fisheries in South Africa

Despite the early attempts by the government to set up catch fisheries in inland waters to benefit rural communities, very few projects were successful [13, 15]. In 1978, there were attempts to develop various gillnet fisheries in Darlington Dam, Eastern Cape, based on gutted *Labeo umbratus* and *Cyprinus carpio* [16]. According to research, the project was initially successful supplying freshly caught fish to various communities in Grahamstown and Uitenhage [15]. However, the project collapsed due to internal conflicts. A similar project was later implemented in Bloemhof Dam in the Free State [17]. There is little information on the attempts to set up similar projects in most of the dams in the country.

However, there have been a few government-sponsored inland fisheries programmes in various places since 1994. A community-driven fisheries project was implemented in the province of the Eastern Cape on the Great Fish River and in two minor impoundments. It was demonstrated that this project helped the local communities generate revenue and provide food security [13]. Another subsistence fishery was established at Ntenetyana dam. Studies conducted at the site suggested that fishing operations could be increased to the point where a small-scale, community-based fishery was created that included both a recreational and hook-and-line subsistence component [17].

There have been several attempts to establish small-commercial fish harvesting at Gariep dam in the Free State. In 2004, the provincial government in the Free State, through the Department of Economic Affairs, Environment and Tourism (DEAET), tried to establish a fishery project for the Venterstad and Oviston communities Venterstad Community Fisheries Project (VCFP) around the dam. The aim was to create economic opportunities for historically-disadvantaged individuals by facilitating access for the communities in designated locations of the lake. The project has not been successful, though. Its failure has been attributed to, among other things, lack of adequate knowledge, lack of consultation, poor planning and management [18].

The aim to promote inland fisheries at Gariep dam can be traced back to the early 1960s when the state constructed a fish hatchery along the Orange River, which supplies the Gariep dam. The hatchery is one of the state hatcheries that were built in the 1960–1970s throughout the country. Thus far, it is the only the hatchery that is fully functional after extensive renovations in 2009 with the help of the Chinese Ministry of Agriculture. However, the focus on the centre now is to produce African Sharptooth catfish fingerlings for aquaculture and not for fisheries purposes. Currently, there is no small-commercial fisheries at the Gariep dam, but only subsistence and recreational fishing is thriving.

4. Small-scale fisheries and subsistence inland fisheries

Small-scale fishing is defined as fishing that uses traditional ways to capture fish for both food and profit. Such operations usually employ few people (±10) for labour. Small-scale fishing operations are frequently family-run and can be distinguished by their labour-intensive nature and low capital investment. Subsistence fishing, on the other hand, is when people fish to satisfy their fundamental needs, such as food. There are few small-scale and numerous subsistence fishing activities taking place in most inland water bodies in almost all inland provinces. In these areas, fishing is mainly for food security and livelihoods for rural communities. However, for a long time, these sectors have primarily remained informal, with no defined governance structures or stakeholder organisations. The limited number of authorised small-scale fisheries utilise basic, manually operated equipment such as treknets or gillnets. In most cases, the economic worth and overall yields are typically unknown. The irony, however, is that most of the subsistence and small-scale fishermen live in disadvantaged areas and lack a steady source of income and are mostly dependent on this activity to provide for their families, which leaves them with little left over to sell for a profit. Therefore, the sector's contribution to the South African economy is minimal. A few factors, such as commerce, cultural preferences, and household poverty levels, are reported to affect the sector's contribution to food security or the economy. In most instances, there is little enhancement in the value of catches, and the fish are typically sold fresh through informal channels or consumed by the household. Fishers have highlighted several factors that they consider as a hindrance to their fishing activities. This encompasses, among other factors, fishing rights, traditional and customary fishing practises, and lack of recognition by the government and other stakeholders.

5. Potential economic benefit of the inland fisheries sector in South Africa

South Africa is one of the African countries where levels of poverty continue rising, with the most recent statistics from 2024 showing a slight increase from previous years. Young people are the most impacted, dealing with issues including underemployment, unemployment, and unethical labour. Although several initiatives, and programmes have been introduced to assist unemployed people, available data shows these are not sufficient to eradicate the challenge of poverty, unemployment and inequality, which are commonly referred to as the "triple challenges" in the country. The problem is severe in rural communities compared to urban areas. This has resulted in many unemployed people resorting to entrepreneurship, with some focusing on aquaculture and fisheries. For many unemployed youth and women, the fisheries sector provides a better alternative worth pursuing because of the low operational costs involved.

Inland fisheries resources in South Africa have the potential to support economic growth, job creation, and food security. However, there lack of research on fish populations and economic potential of fisheries in most of the dams in the country has been identified as one of the most limiting factors affecting the industry. The available data has so far been limited to few small dams and is outdated. These include studies conducted in several small dams in the Eastern Cape [16, 19–22]; Gariep dam and Van der Kloof dam on the Orange River system in the Free State [6, 18, 23–29]; Hartbeespoort Dam on the Crocodile- and Magalies River systems in the North West Province [30].

Dams in the Eastern Cape seem to be most researched for their inland fisheries potential. Among the dams investigated for their fisheries potential in the Eastern Cape in the early 1990s, research concluded that the Xonxa reservoir displayed favourable catch rates for the establishment of a fishery based on *Labeo aeneus* [19]. Later in 2009, Richardson et al. [22] again investigated the stock assessment of the dam, especially the biology of *L. aeneus* and *C. gariepinus*. The aim of the stock assessment models was to assess the sustainable development and management of fisheries project in the dam. Two species, namely *Labeobarbus aeneus* and *Clarias gariepinus* were found to be exploitable for a sustainable fishery. The study indicated that a gillnet fishery for *L. aeneus* (60 mm stretched mesh) could be established (harvest 23 t y^{-1}) while a longline fishery could be possible for *C. gariepinus* (harvest 4 t y^{-1}) [22]. Other dams in the Eastern Cape that were investigated for their fisheries potential include Sinqemeni, Ndlambe, Dimbaza Katriver, and Laing [20, 21]. The research focused on the exploitation of *L. umbratus* populations, which is one of the most occurring species in these dams. The research conducted in these dams concluded that populations of this species would be better suited for exploitation in small, shallow, slightly-enriched reservoirs because *L. umbratus* grows more quickly than populations from more oligotrophic dams [20, 21]. Darlington Dam in Eastern Cape is also among dams investigated for its capacity to sustain an inland fisheries project. *Labeobarbus umbratus*, *C. carpio*, and *C. gariepinus* are the most dominant species at the dam. Research was conducted at the dam to examine the populations of *L. umbratus*, *C. carpio*, and *C. gariepinus* for proposed commercial fisheries (longline and gillnet) and existing recreational angling [16]. Models for stock assessment suggested that gillnet fishing (mesh size of 100 mm) could be implemented. However, until the complete impact of the fishery on the stock is known, the researchers cautioned that annual harvests should not surpass 60 tonnes [16].

The Gariep dam in the Free State is another dam with an economic potential for fisheries. It is the largest dam in South Africa, with a total storage capacity of roughly 5,340,000 megalitres (5340 hm3) and a surface area exceeding 370 square kilometres (140 sq. mi) when at full capacity. Studies conducted in the dam in the 1980s showed a health population of largemouth yellowfish *L. kimberleyensis,* smallmouth yellowfish *L. aeneus*, *L. capensis*, *L. umbratus*, *C. gariepinus* and *C. carpio* [23]. *Cyprinus carpio* was regarded as the species with the greatest harvest potential. The study concluded that a commercial gillnet fishery could be implemented at an annual catch of 886 tons (multi-species). A recent study conducted at the dam by Swanepoel [31] confirmed the potential exploitation of the dam for fisheries. The study concluded that the dam can provide rural communities with economic opportunities, food security and improve livelihoods [31]. This dam is currently supporting several fishing activities at substance level, indicating its potential for an establishment of a small-scale-commercial fisheries enterprises.

In the North-West province, studies for the potential establishment of inland fisheries were conducted in the Hartbeespoort Dam [30]. This dam is dominated by benthic feeding species, namely *C. gariepinus* and *C. carpio*, a condition attributed to the eutrophic state of the dam. The results of the study were not conclusive as to which species had the greatest potential for fisheries [30]. Nevertheless, owing to the ecological state sate of the dam, the study advocates for the harvesting of 200–300 tonnes of these species to facilitate the restoration of zooplankton and macrobenthos communities. The researchers postulated that this would promote the dominance of Mozambique tilapia, *Oreochromis mossambicus*, in the dam. They recommend that a community fishery project using mainly gillnets could be established at the dam. however, there is thriving fisheries projects at the dam.

It is clear from the preceding paragraphs that there are few studies that have been conducted to assess the biological sustainability of harvesting fish from South African impoundments. The lack of information on the population levels and biology of species targeted for fisheries is a major problem hindering the exploitation and consequently also the economics of the inland fishery sector in South Africa. Recently, the South African government, through the Department of Fisheries, Forestry and the Environment, embarked on an initiative to expand and formalise the inland fisheries industry in order to realise its potential to contribute to the economy. New studies are being undertaken at various dams to assess the resource status of inland fisheries in South Africa to assist the government to develop policies that promote and manage the utilisation of the resources in a sustainable manner for the benefit of all communities. These studies show that inland fisheries in South Africa can provide rural communities with economic opportunities, food security and improve livelihoods [31, 32]. South Africa is reported to have more than 700 public dams, most of which have not been explored. Most of these dams and rivers are found in poor rural areas where they are primarily used for domestic and agricultural purposes. Some of the dams with a tremendous potential to sustain a thriving fisheries sector that are currently be assed include the Gariep Dam in the Free State, Vaal dam, Gauteng Province,

Flag Boshielo dam, Limpopo Province, Nandoni dam, Limpopo Province, and the De Hoop Dam, Limpopo Province, Pongolapoort, in KwaZulu-Natal, Voëlvlei, in the Western Cape Province, and Loskop, in Mpumalanga. Most of these dams are already supporting several substance fishery activities, providing a vital source of protein and a source of income for surrounding communities. However, there is no small-scale or commercial inland fishing activities in these dams. Historically, South Africa's commercial inland fishing has been restricted to a small number of dams.

6. Economic potential of substance and small-scale fisheries in rural communities

Since the small-scale fishing industry is mostly an informal activity without a formal framework for stakeholder participation or data collection, there is a dearth of published information on the scope and livelihood significance of this activity in rural communities. To describe the existing small-scale fishery and evaluate the relevance of indigenous knowledge in inland fisheries, case studies of a few chosen fishing communities were undertaken. The findings showed that small-scale fishing activities for livelihood purposes was present on 77% of inland water bodies in South Africa. These investigations also discovered that there was minimal indication of value addition and that freshwater fish value chains were brief. Usually, caught fish are for family consumption or sold fresh at nearby busy intersections. A study conducted using a GIS model to identify regions of higher potential based on the relationships between climate, geography and fish yield to predict areas showed that large South African impoundments have a fishery production potential of 15,000 t. However, this production potential is good for subsentence and small-scale fisheries, but it is not feasible for the establishment of large-scale commercial or industrial fisheries in inland waters. Therefore, the best way to use inland fisheries for livelihoods is for recreational and small-scale subsistence fishing. According to consultations with small-scale fishing communities, small-scale fishing is a significant source of income for many households and should be acknowledged and supported by inland fishing policies.

7. Current challenges

The studies conducted at some dams in South Africa have identified some of the challenges affecting the growth and sustainability of the inland fisheries industry. Apart from the lack of research data on the sustainability of fisheries at most dams, researchers have highlighted the lack of laws and policies as a challenge to the successful development of the sector. Laws and policies are needed to regularise the industry and development management plans for fisheries activities. A detailed list of specific challenges affecting the sector in South Africa include poor governance, inadequate organisational structures, restricted access to markets, social services, and financial resources, low participation of small-scale or subsistence fishermen in decision-making, overfishing, poor fishing methods, and a lack of value-adding mechanisms. Researchers recommend that government needs to undertake fundamental reforms to improve the sector's governance and develop policies that will recognise, protect, manage and promote inland fisheries. They further argue that new strategies focussing on marketing and building of processing plants are critical as this will add value to catches and help rural households make more profit.

8. Policy and legislation

South Africa has had no inland fisheries policy until recently (2021). The lack of the policy has resulted in the sector not receiving recognition and not being managed effectively. While the government has so far effectively and efficiently governed other public resources including land, water, minerals, and sea fisheries, inland fisheries were largely ignored. Experts and fishermen have listed the lack of a guiding policy/legislation as the main factor affecting the development of the inland fisheries in South Africa. In many places fishing in dams and rivers has been regarded as illegal activity and the lack of regulation resulted in serious problems for the industry such as overfishing or arrest of fishers by authorities. The lack of a fair governance policy has also resulted in the proliferation of using unregulated fishing practises. In South Africa, existing policies regarding inland fisheries have mainly focused on conservation and biodiversity rather on the sector as a source of income, food security, or economic contribution. As a result, several small-scale fishermen have voiced their worries that the government and other stakeholders do not acknowledge their fishing rights, traditional and customary fishing methods, or the sector's contribution to the upliftment of rural lives.

To address this problem, DFFE recently embarked on the initiative of developing an inland fisheries policy. The goal of this policy is to bring inland fisheries governance into compliance with constitutional mandates for a sustainable development strategy that utilises natural resources for the good of all residents. Although gazetted and promulgated in 2021, the full implementation and effect of the policy has not yet had an impact on the sector. Clearly, the implementation of the policy will take time to have effect due to the diverse inland fisheries resources. The new policy guarantees a comprehensive, multi-departmental, and multi-stakeholder approach to the sustainable growth of the inland fisheries sector. It also tackles the present necessity for transformation and expansion of value chains associated with the inland fishing industry. Henceforth, the informal and unrecognised activities of small-scale fishers in inland areas would be formalised and regulated according to national legislation. Additionally, in contrast to the current situation, where fishing activities are regulated by the provincial departments responsible for environmental management, the department (DFFE) will promulgate national and provincial legislation that permits and authorises the issuance of permits and authorisations to individuals, legal entities, or community groups. The department has suggested that, by considering ecological constraints like waste management, pollution reduction, habitat preservation and restoration, and sustainable harvesting of fisheries resources, the policy would enable the fisheries sector to develop sustainably and contribute more to economy. The policy promises to give priority to subsistence and small-scale fishermen when granting permits. Among other things, the policy promises to guarantee that all resource-users will priority have access to an effective and user-friendly registration and permitting system.

9. Important consideration-what must be done to promote rural inland fisheries in SA?

Most of the dams identified to have the greatest potential for inland fisheries were built in rural communities. Therefore, the sector should be improved so that it can used to uplift poor communities from poverty. The question, "what must be done to promote rural inland fisheries in SA" represents an important step towards the realisation of the economic potential of inland fisheries for rural communities in South Africa. It sets the tone and allow the government to set priorities about what needs to do urgently address or remove all obstacles. Despite the government having achieved one of the most important requirements (enacting an inland fisheries policy which scientists have been consistently calling for), several other issues require urgent attention before the full potential of the sector can be realised. Without them being addressed, the policy will not achieve its intended purpose. To address some of these challenges, government must develop and implement fisheries management plans to ensure the long-term sustainability of the fisheries sector. Extension officers will have to be trained to educate communities and fishermen about the implementation and benefit of the policy in addressing overfishing while improving access to fisheries resource. These interventions will address concerns of fairness and capability in order for communities to take full advantage of inland fisheries-based livelihood opportunities. It is also necessary for inland fisheries to be legally recognised for their socioeconomic benefits and for supporting rural livelihoods.

More research is still needed to assess the inland potential of most dams that are currently being used for subsistence and small-scale fisheries. The dearth of information regarding the sector's productivity, sustainability, and possible effects on the biodiversity of indigenous species is one of the main obstacles hindering its growth on the majority of South African water bodies. Therefore, research surveys and stock assessments should be prioritised. Other considerations that research has to address include climate change adaptation. Climate change has the potential to alter the future of inland fisheries. Investigating and developing measures to adapt to climate change will guarantee the long-term sustainability of the sector.

Marketing of catches from subsistence fisheries is also limiting factor affecting the economic potential of the sector, especially at subsistence level. The majority of the produce is used for domestic purposes, with the remainder being sold to neighbours or unofficial marketplaces without any value-addition, which means minimal profit. It is thus critical for government to invest in infrastructure that will promote inland fisheries development in rural communities and ensure they benefit from fisheries resources. These will include, among others, investing in roads, storage facilities for caught fish, processing plants, value addition and marketing. In most cases, the fish is sold fresh, which means it cannot be kept for long, as it will start spoiling. Finally, it is imperative for government, through partnerships with the private sector, to create financial support platforms designed to support rural communities to invest in inland fisheries enterprises and to promote access to markets for fish products. Therefore, it is important for the government to prioritise public sector initiatives that raise the value of fish products from fishing communities.

Author details

Esau Matthews Mbokane[1*] and Hlulani Archiebold Hlungwani[2,3,4]

1 Aquaculture Research Unit, School of Agricultural and Environmental Sciences, University of Limpopo (Turfloop Campus), Sovenga, South Africa

2 Water Research Commission, Gezina, South Africa

3 Department of Biodiversity, School of Molecular and Life Sciences, University of Limpopo (Turfloop Campus), Sovenga, South Africa

4 Sustainable Aquaculture Management, The Department of Forestry, Fisheries and the Environment, Cape Town, South Africa

*Address all correspondence to: esaumbokane@yahoo.com

References

[1] FAO. Review of the state of the world fishery resources: Inland fisheries. In: FAO Fisheries Circular. No. 942, Rev.1. Rome: FAO Inland Water Resources and Aquaculture Service, Fishery Resources Division; 2003. p. 60

[2] Marshall B, Maes M. Small Water Bodies and their Fisheries in Southern Africa. CIFA Technical Paper No. 29. Rome: FAO; 1994. p. 68

[3] Neiland AE, Chimatiro S, Khalifa U, Ladu BMB, Nyeko D. Inland Fisheries in Africa: Key Issues and Future Investment Opportunities for Sustainable Development. Technical Review Paper - Inland Fisheries. NEPAD; 2005. p. 76. Available from: www.fishforall.org/ffasummit/C_Eg/Inland%20Fisheries%20Review%20Paper

[4] Andrew TG. Final Report on the Activities of the Rural Fisheries Programme, 1997-2000. Rural Fisheries Programme, Department of Ichthyology and Fisheries Science. South Africa: Rhodes University; 2001. p. 155

[5] Weyl OLF, Potts WM, Rouhani Q, Britz PJ. The need for an inland fisheries policy in South Africa: A case study on the Northwest Province. Water SA. 2007;**33**:497-504

[6] Bok A. The demography, breeding biology and management of 2 mullet species (Pisces: Mugilidae) in the Eastern Cape, South Africa [Ph.D. thesis]. South Africa: Rhodes University; 1983. p. 268

[7] Van den Berg RA, Gaigher IG, Lenyai SK. The use of small farm dams in the Transkei and Ciskei for the production of table-fish using Aischgrund carp. Journal of the Limnological Society of Southern Africa. 1975;**1**:11-15. DOI: 10.1080/03779688.1975.9632902

[8] Roode MC. Harvesting of freshwater fish from natural resources: A marketing experiment conducted in the Transkei. Fish Farmer Transvaal Newsletter. 1978;**20**:2-12

[9] Van der Waal BCW. Status of fisheries of black states of South Africa. In: Collected Papers from the Symposium on Fisheries and Rural Development; 20-21 November 1978. Grahamstown, South Africa: Rhodes University; 1978. pp. 171-179

[10] Mabitsela CK. An ecological investigation of the economically viable fish species in certain dams in Lebowa, South Africa [M.Sc. thesis]. South Africa: University of the North; 1981

[11] Saayman JE, Jooste A, Olivier PAS, Mokgalong N, Polling L. The cropping potential of Nzhelele Dam, Republic of Venda, on self-sustained basis. Report of the Department of Zoology. South Africa: University of the North; 1983. p. 12

[12] Schoonbee HJ, Hecht T, Saayman JE, Polling L. Fish population assessment in a temperate Lowveld impoundment of the Transvaal, South Africa. Water SA. 1995;**21**(2):147-152

[13] Andrew TG, Rouhani QA, Seti S. Can small-scale fisheries contribute to poverty alleviation in traditionally non-angling communities in South Africa? African Journal of Aquatic Science. 2000;**25**:49-55. DOI: 10.2989/16085914.2020.1745413

[14] Rouhani Q, Britz PJ. Contribution of Aquaculture to Rural Livelihoods in

South Africa: A Baseline Study. WRC Report No. TT 235/04. Pretoria: Water Research Commission; 2004. p. 105

[15] McCafferty JR, Ellender BR, Weyl OLF, Britz PJ. The use of water resources for inland fisheries in South Africa. Water SA. 2012;**38**(2):327-344. DOI: 10.4314/wsa.v38i2.18

[16] Weyl OLF, Booth AJ, Winker H, Traas GRL, McCafferty JR, Peel RA, et al. Assessment of the Fishery Potential for Alien Species on Darlington Dam. Report No. 5. SAN-Parks: Sundays River Freshwater Fish Research Project, Project ID: PO64438; 2010. pp. 1-59

[17] Rouhani Q. Report for the Assessment of the Potential for Community based Small-Scale Fisheries in Ntenetyana Dam (Alfred Nzo District Municipality). Rural Fisheries Programme, Department of Ichthyology and Fisheries Science. South Africa: Rhodes University; 2003. p. 12

[18] Potts WM, Weyl OLF, Andrew TG. Final report on an initial assessment of the fisheries potential of Gariep Dam Report for: Department of Economic Affairs, Environment and Tourism, Northern Region, Province of the Eastern Cape. South Africa: Enviro-Fish Africa (Pty) Ltd; 2004. pp. 38

[19] Schramm M. A comparison of Cyprinid fish populations in three large Transkei reservoirs with respect to abundance and reproduction. African Journal of Aquatic Science. 1993;**19**:74-88. DOI: 10.1080/10183469.1993.9631340

[20] Potts WM. Towards the development of species-specific fish production models for small reservoirs in Southern Africa [Ph.D. thesis]. South Africa: Rhodes University; 2003. p. 148

[21] Potts WM, Booth AJ, Hecht T, Andrew TG. The life history and fishery potential of *Labeo umbratus* (Teleostei: Cyprinidae) in five small South African reservoirs. African Journal of Aquatic Science. 2006;**31**(2):285-295. DOI: 10.2989/16085910609503898

[22] Richardson TJ, Booth AJ, Weyl OLF. Rapid biological assessment of the fishery potential of Xonxa Dam, near Queenstown, South Africa. African Journal of Aquatic Science. 2009;**34**:87-96. DOI: 10.2989/AJAS.2009.34.1.9.734

[23] Hamman KCD. Aspekte van die bevolkingsdinamika van die Hendrik Verwoerddam met verwysing na die ontwikkeling van 'n visserybestuurplan [Ph.D. thesis]. South Africa: Rand Afrikaans University; 1981. p. 208

[24] Allanson BR, Jackson PBN. Limnology and Fisheries Potential of Lake le Roux. South African National Scientific Programmes Report No. 77. Pretoria: CSIR; 1983. p. 182

[25] Tómasson T. The biology and management consideration of abundant large cyprinids in Lake le Roux, Orange River, South Africa [Ph.D. thesis]. South Africa: Rhodes University; 1983. p. 218

[26] Tómasson T, Bruton MN, Hamman KC. The demography and management of large Cyprinids in a reservoir on the Orange River, South Africa. Fisheries Research. 1985;**3**:279-308. DOI: 10.1016/0165-7836(85)90028-1

[27] Ellender BR, Weyl OLF, Winker H. Who uses the fishery resources in South Africa's largest impoundment? Characterising subsistence and recreational sectors on Lake Gariep. Water SA. 2009;**35**:677-684. DOI: 10.4314/wsa.v35i5.49194

[28] Ellender BR, Weyl OLF, Winker H, Booth AJ. Quantifying the annual fish harvest from South Africa's largest

freshwater reservoir. Water SA. 2010;**36**(1):45-52. DOI: 10.4314/wsa.v36i1.50906

[29] Ellender BR, Weyl OLF, Winker H, Stelzhammer H, Traas GRL. Estimating angling effort and participation in a multi-user, inland fishery in South Africa. Fisheries Management and Ecology. 2010;**17**(1):19-27. DOI: 10.1111/j.1365-2400.2009.00708.x

[30] Koekemoer JH, Steyn GJ. Fish Community Study of Hartbeespoort Dam. Final Report. North West Province, South Africa: Department of Agriculture, Conservation, Environment and Tourism; 2005. p. 101

[31] Swanepoel PJ. Assessing Small-Scale Fisheries Potential of Gariep Dam, South Africa's Largest Impoundment. 2022. Available from: http://hdl.handle.net/11660/12083

[32] Hlungwani HA. Assessing the Potential of Establishing a Freshwater Inland Fishery in Flag Boshielo Sam, Limpopo Province, South Africa2020. p. 180

Chapter 6

The Importance of Plankton in Marine Fish and Shellfish Larval Nutrition

Mohamed Ashour, Einar Ringø, Ehab El-Haron and Ashraf M.A. Goda

Abstract

Plankton, comprising phytoplankton and zooplankton, are the primary food sources for marine fish and shellfish in their early larval stages, supplying essential nutrients and energy for growth and development. This chapter provides an overview of the vital role of plankton in the nutrition of marine larvae in hatcheries. This chapter explores the nutritional and biochemical composition of plankton, stressing their high quantities of proteins, lipids, vitamins, minerals, and essential fatty acids, all of which are very critical for marine larval stages. This chapter addresses the specific dietary requirements of fish and shellfish larvae, highlighting the importance of plankton diets in larval development and survival. This chapter also looks at the challenges and considerations associated with producing and utilizing phytoplankton and zooplankton in marine hatcheries. This chapter emphasizes the importance of plankton as a crucial component of marine larval feeding, hence helping to the success of aquaculture production and wild conservation.

Keywords: marine larval rearing, shrimp, diatoms, green water, rotifer, copepoda

1. Introduction

As the world's population has increased, aquaculture has grown fast [1]. Feeding approaches for target aquatic species from larval to fully grown adults in both freshwater and marine aquatic animals are essential for effective aquaculture [2]. Because of their youthful digestive tracts and lack of vital digestive enzymes, aquatic organisms have a natural predilection for live feeds like plankton (phyto- and zooplankton). All aquatic larvae, in particular, are entirely dependent on live food [3]. One of the most important challenges in fresh and marine aquaculture is the availability of live feed (both in quality and in quantity). Aquaculture's developing infrastructure is already grappling with a number of difficulties, including inadequate nutrition, high larval mortality rates, and sexually immature adults. All of these issues are linked to their diet. Notably, because they are sensitive to aquatic physicochemical factors and diets, substantial mortality occurs during the post-hatching and yolk absorption periods. They are unable to consume substantial, robust natural, or artificial food because of their toothless, tiny mouth [4].

To digest food, aquatic larval stages require appropriate carbohydrates, lipids, proteins, vitamins, minerals, and enzymes. Without these components, it is impossible to manage broodstock [5, 6]. Aquatic animal larvae (fish and crustaceans) feed on their yolk sac or an external microalgal diet for the first 2 weeks before transitioning to live zooplankton, such as copepods, rotifers, cladocerans, and crustaceans [7]. The stomach is absent in newborn larvae, and the hindgut's epithelial cells must have the maximum space for protein digestion [8]. Because of their superior performance, live feeds are always preferable for larvae as they are easier to digest than inert formulated feeds; even modified artificial feeds cannot be compared to live feeds [9]. According to their taste and mouth gap, larvae prefer live prey. All aquatic organisms are active feeders who prefer to consume live prey; in particular, the prey's movements—such as swimming, whelming, and jerking—influence the frequency and responsiveness of feedings [10]. Furthermore, as active predators, larvae always prefer chasing and consuming visually enticing moving zooplankton to immobile algae or inert artificial pellet feeds [11]. Because of their immature digestive tracts, larvae require easily digestible diets that contain crucial digestive enzymes. Only live zooplankton may meet this special requirement [12].

Larvae require basic nutritional requirements such as proteins, lipids, carbohydrates, vitamins, and minerals for proper development [13]. As a result, live food and zooplanktons are more appealing to larvae compared to prepared dry and inert meals due to their slim exoskeleton and high water content (more than 80%). Despite this, some phytoplankton, in particular, are harmful to digestion due to their stiff cell walls [14]. According to Kandathil Radhakrishnan [15], omnivorous and carnivorous species prefer zooplanktons, primarily rotifers, copepods, cladocerans, and crustaceans, as live prey from the larval stage to adulthood. Zooplanktons mostly consume bacteria and algae to obtain their vital nutrients [16]. Therefore, these secondary grazers rely on aquatic algae as their primary supply of nutrients. Linoleic acid and linolenic acid are the chemical names for the important fatty acids (FAs) [16]. Most significantly, algae are the only source of these linoleic and linolenic acids. Therefore, zooplankton either directly needs vital fatty acids from algae or converts them from linolenic acid. Therefore, zooplankton provides the higher trophic level with essential nourishment [17].

Many freshwater fish, marine and coastal fish, crustaceans, and mollusks require fatty acid-rich diets to grow properly. The shift from embryonic to juvenile stage is characterized by frequent changes in living habits and fast transformation [18]. In hatcheries, for fish and crustacean larvae, a particular diet that is sufficient in nutrients is required during this voyage. Furthermore, brood organisms require a consistent supply of nutrients from wholesome food [18]. Feed conversion, fertility, growth, molting, egg hatchability, resistance to osmotic stress, and brood survival are all impacted by the nutritional value of live or artificial feed. This chapter will provide an overview of the important role of plankton in the nutrition of marine larvae in marine hatcheries. The present chapter discusses the specific dietary requirements of fish and shellfish larvae, emphasizing the importance of plankton diets for larval development and survival. Moreover, this chapter investigates the challenges and considerations associated with phytoplankton and zooplankton production and utilization in marine hatcheries, which contribute to the success of aquaculture production and conservation in the wild.

2. Dietary requirements of marine fish and shellfish larvae

During their early stages of development, marine fish and shellfish larvae are extremely susceptible and require specific biotic and abiotic factors stimuli to, survive, develop, and thrive [19]. Numerous studies and reviews address various facets of larval nutrition and illustrate achievement in understanding from several points of view. Although nature provides a wide range of conditions for developing larvae, laboratory research employing simplified techniques in controlled environments with limited prey options and stable abiotic and biotic factors is largely responsible for our understanding of nutrition during the early stages [20]. The differences in feeding physiology and nutritional demands between species including those within the same family should be recognized. As a result, numerous particulars are needed [21]. Consequently, many individual systems require specialized research and cannot be easily generalized from findings on model species [22]. To increase the quality of aquatic larvae and juveniles, it is understood that a thorough understanding of the nutritional needs of larvae throughout their development will help to optimize feeding procedures [23]. However, given the fragility of marine larvae, it is mainly hard to evaluate and provide nutritional needs when multiple metabolic and physiological limitations are interconnected and may potentially impede growth or proper development [24]. Developing diets for marine larvae that satisfy their needs in order to ensure the best possible ingestion, digesting, and absorption requires a full knowledge of the various processes and components involved in food acquisition and processing [25]. However, the most important dietary requirements of marine fish and shellfish larvae are (1) protein content and related amino acids profile, (2) lipid content and related fatty acids profile, (3) macronutrient, (4) minerals, and (5) vitamin [26].

2.1 Lipid content and fatty acids profile

A considerable information gap regarding the food needs of fish and shellfish larvae and early juvenile stages impedes aquaculture development. According to growth and survival investigations, the success of these early fish and shellfish larval stages is greatly determined by the nutritional quality of starting live prey, which are often composed at least in part of enriched live prey, and the optimization of feeding conditions [27]. According to Glencross [27], lipids are an important dietary component and the primary source of metabolic energy that supports the early stages' rapid development. Compared to proteins or carbohydrates, lipids offer at least two-three quarters more energy per gram [28]. Neutral lipids, including free fatty acids, sterols, and triglycerides, have a quick turnover and can satisfy short-term energy needs; however, the β-oxidation of saturated fatty acids (SFAs) releases energy more effectively than polyunsaturated fatty acids (PUFAs) [29]. Numerous research have been undertaken on fish larvae's lipid requirements, which include both the ratio of phospholipids to neutral lipids and necessary fatty acids. Nonetheless, very few studies have attempted to determine the quantitative requirements for these nutrients with dose response [26, 29].

Lipids continue to be the most poorly understood nutrient following years of research [27]. Because they cannot be biosynthesized to sustain normal development, certain FAs are regarded as EFAs for the development of marine fish [22]. Because EFAs are involved in many physiological processes, including immunity, ion balance regulation, muscular contraction, cell adhesion, buoyancy control, and brain and eye development, they directly affect the growth and survival of marine larvae [30].

The majority of these EFAs are derived from two closely related families (n-3 and n-6), as well as three long-chain polyunsaturated fatty acids (PUFAs) that have long been regarded as essential: docosahexaenoic acid (DHA, 22:6n-3), eicosapentaenoic acid (EPA, 20:5n-3), and arachidonic acid (ARA, 20:4n-6). While ARA and EPA are precursors of bioactive eicosanoids, DHA and EPA play significant roles in maintaining membrane fluidity [31]. DHA is preferentially absorbed into retinal and neurological tissue during larval development, and low DHA levels result in growth defects and increased mortality. Fish require dietary fatty acids, including linoleic acid (18:2n-6) and α-linolenic acid (18:3n-3), which are stored in muscle tissue for physiological demands. Although these FAs are precursors of DHA, EPA, and ARA, most marine species have low activity of the particular enzymes that are responsible for their production [32].

Artemia and rotifers are the principal live prey employed in aquaculture and they contain significant levels of phospholipid (PL), which is important for fish nutrition [33]. They must be enriched before being utilized as prey, though, because their low levels of EFA are insufficient for the early stages of life [34]. Copepods, on the other hand, are abundant in phospholipids and EFA and are naturally occurring fish prey in the wild, nevertheless, producing them in the laboratory is tough and complex [35]. Enrichments can be used to change the fatty acid composition of rotifers and *Artemia*, even while their PL content per dry weight (DW) cannot be changed [36]. Since copepods and other natural prey contain a high concentration of PUFA, the influence of the type of FAs binding to the PL has also been discussed. For example, PL from bonito eggs enriched in n-3 highly unsaturated fatty acids enhances growth and survival in larvae of *Plecoglossus altivelis* more effectively than PL from vegetable sources [37]. The effects of EFAs on the growth, behavior, survival, and biological processes of marine fish larvae have been extensively studied; however, few studies have quantified the requirements in developing larvae and in distinct species. It is important to remember that the proportional importance of each FA varies by species [38].

2.2 Protein content and amino acids profile

The specific types of protein and amino acids (AAs) in the diet have an important role in the life cycle of fish and shellfish larvae [39]. As presented in **Table 1**, adding moderate amounts of protein throughout the life cycle of fish and shellfish species has been proven to improve their growth and survival rates [40]. The overall AA composition of freshly hatched marine fish eggs ranges from 40 to 60% of their dry mass and comprise proteins, peptides, and free AAs [41]. These AAs are derived from yolk protein. Free AAs are an important source of energy for marine fish embryos before the digestive system of the hatched larva develops sufficiently to begin external eating [42]. For example, during the egg and yolk-sac larval stages, the amount of free AAs in Atlantic cod (*Gadus morhua*) eggs decreased from 200 nmol/egg at spawning to 25 nmol/egg or fish. Free AAs are key substrates for aerobic adenosine triphosphate (ATP) production in eggs and yolk-sac larvae; therefore, an adequate supply of them is required for normal development of the embryo [43].

A balanced food provision in broodstock diets can promote fecundity or egg quality by affecting the brain-pituitary-gonad-endocrine system or the accessibility of a chemical required for egg production [43]. A high protein diet can improve the Nile tilapia's total AA pool, fertilization, and hatchability of their eggs. Notably, when fish were given a broodstock diet that included only 10% crude protein, the females'

Stages/phases			Feeding items	The main functions of amino acids
Fish	Shellfish			
	Shrimp	**Crab**		
Fertilized egg	Fertilized egg	Fertilized egg	—	Improves and enhances the quality of fertilized eggs
Embryos	Nauplius	—	Yolk stocks	Improves and enhances embryos development
—	Zoea	Zoea	Microalgae	Enhances survival and growth development
—	Mysis	Megalopa	Phyto/zooplankton	
Larvae	—	—	Phyto/zooplankton	
Fry	Post-larvae		Zooplankton/ microdiets	
Fingerling	Juvenile	Juvenile	Diets/Pellet	Enhances survival and growth development
Adult	Adult	Adult	Diets/Pellet	Enhances the development of gonads, sperm, as well as egg production

Table 1.
Essential roles of protein and amino acids during the life cycle of fish and shellfish.

eggs did not fertilize [44]. Several studies have reported that adding enough protein or AAs to broodstock meals improved fertilization, hatchability, and larval growth [42, 45, 46]. When the nutrients in the yolk are no longer sufficient to support the larvae's metabolic requirements, fish and shellfish must begin external feeding [47]. After the first feeding, AAs are also crucial catabolic substrates and can provide up to 60% of the energy needed [23]. Furthermore, because fish and shellfish larvae grow so quickly, they need a lot of dietary AAs to sustain protein synthesis and accretion [48]. As early-stage larvae progress toward metamorphosis, their extracellular proteolytic capacity increases. As a result, free AAs or protein hydrolysates (predigested protein supply) are necessary dietary components to initiate the feeding process [41]. In order to start feeding marine fish larvae, free amino acids (AAs) or protein hydrolysates (a predigested protein source) are crucial dietary components. Dietary supplementation with specific AAs, including taurine, methionine, lysine, and tryptophan, improved larval growth, digestive system development, and metamorphosis [49].

2.3 Macronutrients

In general, the constant feeding during larval stages may affect the optimal macronutrient balance by affecting food availability and gut passage time [50]. The ideal composition is also influenced by the way nutrients are provided Understanding these difficulties is an ongoing process in our science [51]. Utilizing live feed presents difficulties for investigations into the optimal macronutrient composition for fish and shellfish larvae because of the metabolic processes and nutritional content of the feed organism. Results may become more complicated if some diets contain nutrient deficits. Considering these challenges, the research conducted by Morais et al. [52] revealed that unenriched *Artemia* increased growth in one case and indicated a trend toward improved growth in another. Since it was improbable that non-enrichment would positively alter the fatty acid composition of *Artemia*, this outcome was most likely brought on by the unenriched *Artemia's* higher protein-to-lipid ratio [53]. The poor acceptability of most inert diets, particularly semipurified ones, by many fish and shellfish species, makes the use of experimental microdiets a lot harder. Efforts have been made to solve this issue, in spite of these obstacles. When Senegalese sole larvae were fed microdiets with different protein contents, Yufera et al. [54] displayed that a higher protein content led to a slightly better rate of growth and survival as well as a faster pace of eye migration. Nevertheless, there are currently inadequate detailed dose-response studies evaluating more than two levels of macronutrient variation in fish larvae. On the other hand, by experimenting with various dietary compositions, experimental microdiets offer a chance to investigate possible macronutrient preferences. Studies on gilthead seabream larvae that utilized marked food microparticles were inconclusive, implying that the ability to select macronutrients may not be fully established at the larval stage, even if juvenile and adult fish may select appropriate compositions depending on their macronutrient requirements. These findings may be connected to gut development and dietary choices [55].

2.4 Minerals

The studies on the mineral needs of fish and shellfish larvae only began after 2005, and there aren't many publications to date. Since there are just a few species or none at all to measure each nutrient, direct measurements of mineral requirements in fish larvae are dispersed and limited [56]. Furthermore, the presence of minerals in seawater, as well as the likelihood of varied bioavailabilities for distinct minerals, complicates studies on mineral needs. Therefore, further research is required to determine whether the mineral requirements of fish and shellfish larvae differ from those of juveniles and adults [57]. *Artemia* from the Great Salt Lake in the USA demonstrated high mineral content, except for iodine and zinc. The insufficient levels of certain minerals in live prey prompted investigations into the adequacy of mineral content in feed [58]. Nguyen et al. [59] investigated the enrichment of *Artemia* with zinc and manganese. Increasing the dietary Mn (DM) content from 12 to 40 mg/kg of DM led to significant growth in red sea bream larvae, advancing from 15 to 30 days post-hatch (dph). The inclusion of Mn, Zn, or both led to a decrease in skeletal deformities, reducing the percentage of deformed fish from 53% in the control group to 39–41% in the treated groups. Additionally, in an experiment involving cod and utilizing sodium iodide (NaI)-enriched rotifers, efforts were made to maintain high iodine concentrations until feeding the rotifers to the larvae. The control rotifers seemed to possess sufficient iodine (0.6 g/kg) to meet the larval requirements [60].

Conversely, Ribeiro et al. [61] identified reduced growth and alterations in thyroid hormone metabolism in *Senegalese sole* larvae fed on control rotifers and *Artemia*, in contrast to larvae nurtured on iodine-enriched live feeds. Hamre et al. [62] showed that rotifers have a lower mineral content than copepods. Selenium levels in rotifers fell below the National Research Council's (NRC's) [63] standards for fish and shellfish. Furthermore, cod larvae fed on copepods from a pond in Northern Norway demonstrated much greater mineral levels than those fed on rotifers [64]. In a preliminary investigation, Hamre et al. [65] fed cod larvae with iodine- and selenium-enriched rotifers, noting a significantly improved survival rate compared to the control group. While the added selenium was successfully transferred to the larvae, iodine was not, likely due to the poor retention of NaI in rotifers during storage. In their study, Penglase and colleagues [66] fed cod larvae with selenium-enriched rotifers containing up to 3 mg/kg of selenium. They observed minimal impacts on growth and survival, yet noted enhanced gene expression and activity of glutathione peroxidases due to the enrichment, suggesting a need for selenium levels higher than those in the control samples. Selenium concentrations in rotifers ranged from 0.04 mg/kg dry weight to 0.7 mg/kg in the control group. The recognized selenium requirement for juvenile and adult fish typically hovers around 0.35 mg/kg [67].

2.5 Vitamins

The little available data show that fish and shellfish use vitamins differently before their initial feeding [66]. Assessing nutrient use from the yolk provides a useful indication of the necessary nutrients needed for organism development [68]. However, this strategy does not account for potential restrictions caused by inadequate nutrient digestion and absorption. A more thorough examination of the available literature data, including recalculating the yolk intake rate per unit of fish larval growth, can yield insights. These recalculated statistics can be compared to requirements derived using direct methods. Rønnestad et al. [69] observed that the levels of ascorbic acid and α-tocopherol (α-TOH) in entire larvae remained steady during the yolk-sac stage, indicating no loss or utilization of these vitamins. At hatching, a significant portion of ascorbic acid (about 80%) and α-TOH (around 97%) was retained within the yolk-sac compartment. Over time, the levels of ascorbic acid and α-TOH in the yolk diminished, albeit at varying rates. By the first feeding, over 95% of ascorbic acid had been transferred to the larval body, while less than 30% of α-TOH present in the yolk at hatching had been absorbed. The transfer of α-TOH was completed upon complete absorption of the yolk. Because of the many types of vitamin A confirmed in the yolk and larvae, as well as its carotenoid precursors, the vitamin A mass budget during endogenous feeding in halibut is complex [70]. Research on the conversion of these precursors to vitamin A in juveniles has shown that different carotenoids have variable conversion efficiencies. It is unknown how well this conversion works during the critical period before the first feeding when the eyes start to function. Research on B6 demonstrated that approximately 25% of B6 in the yolk of halibut was lost at first feeding. The consumption of B6 was initially slow but enhanced steadily during the yolk-sac stage as the embryo's dry matter content expanded. Rainbow fish showed similar declines in B6 levels during endogenous feeding, with a 15% reduction [71].

3. Plankton production keys

The Chinese saying "Big fish eat small fish, small fish eat shrimp, shrimp eat mud" emphasizes the importance of plankton (microalgae and zooplankton) in the food pyramid and its potential as a therapeutic element. Microalgae form the base of the marine food chain. Microalgae are critical in the commercial rearing of a wide range of marine organism species because they offer food for all stages of bivalve and mollusk growth, larval stages of some crab species, and very early growth stages of some fish species. In addition, microalgae create huge amounts of zooplankton (rotifers, copepods, and brine shrimp), which are used as food for fish and crustacean larvae and early juvenile stages [72], as illustrated in **Figure 1**.

Despite hundreds of species being examined, only about 10 plankton species are commonly employed as live food in marine hatcheries. To be an effective aquaculture species, the correct plankton species must have a number of key qualities [73]. Phytoplankton should have the following basic characteristics: (1) being of the right

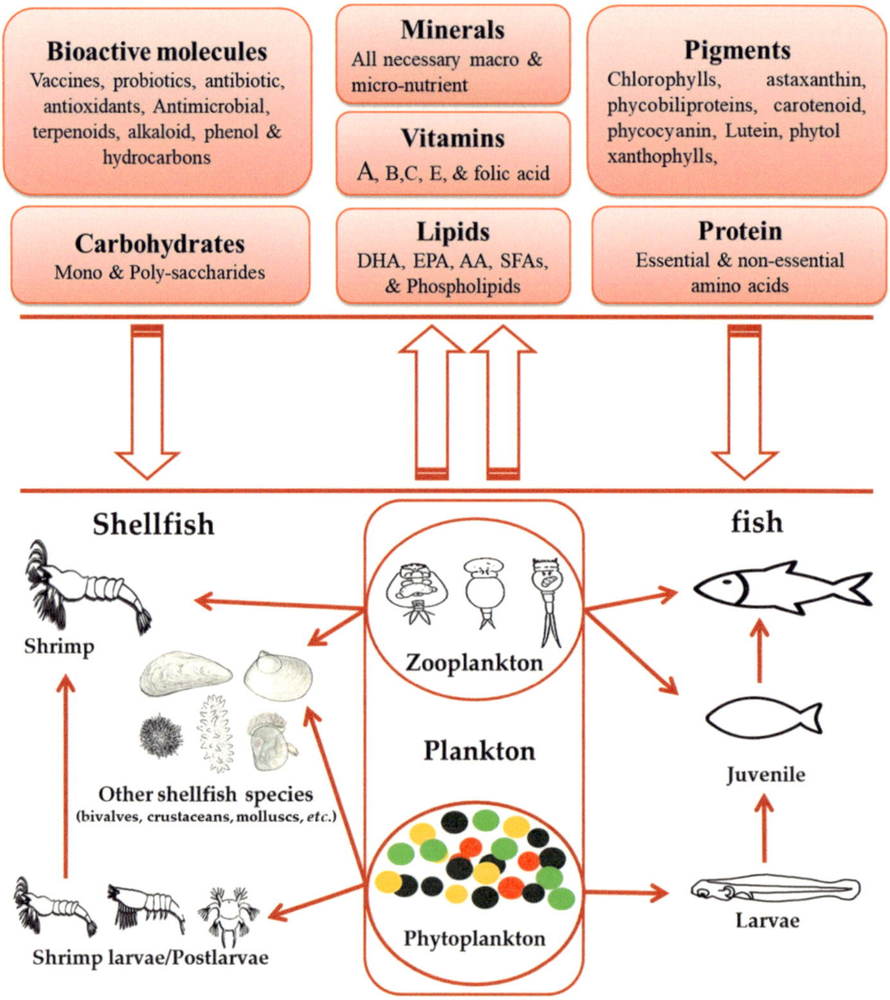

Figure 1.
The role of plankton (phytoplankton and zooplankton) in fish and shellfish larval nutrition, aquatic systems, and aquaculture, in general. Modified from Ashour [72], Elshobary and Ashour [73], and Mansour et al. [74].

size for consumption (between 1 and 15 μm for filter feeders and between 10 and 100 μm for grazers), (2) being easily digested; (3) growing quickly, (4) being simple and stable to mass culture, (5) being adapted to the local environment, (6) being cultivated under nutrient limitation, and (7) having a good nutrient composition [75].

All aquatic ecosystems (fish, shellfish, and zooplankton) benefit from microalgae as the best source of protein, fat, phospholipids, carbohydrates, minerals, vitamins, and the bioactive substances [74]. Microalgae also contain a variety of pigments and bioactive molecules that collectively contribute to a range of biological activities, including probiotics, antibiotics, antimicrobials, antioxidants, and vaccines [76]. Microalgae's biochemical makeup differs within classes and even species. The main organic components are always protein and fat, accounting for 15–70% of dry weight. Depending on nutrient restrictions and growth conditions, the organic and inorganic components of microalgae species vary greatly within a strain [77]. Enhancing the production of excellent nutritional quality and quantity of cultured microalgae is crucial, since live microalgal species are traditionally used as feed for a variety of aquatic organisms in aquaculture. Because it involves major risk variables that make it non-impregnable, microalgae production is one of the most crucial aspects of marine hatcheries. Over 50% of the expenses associated with producing marine larvae are related to the production of microalgae as live food in marine hatcheries [78]. Microparticulate diets, micro-encapsulated and inert food, yeasts, microalgae paste, microalgae free-lipid, and microalgae concentrate preserved by various techniques are just a few of the food sources that have been the subject of numerous studies in recent decades due to these issues [73, 74, 79].

4. Importance and nutritional value of plankton in marine hatcheries

4.1 Phytoplankton

Plankton culture (both phyto and zoo) is an important aspect of hatchery activity [80]. On the other hand, plankton production is often managed separately from aquaculture on a global scale. Numerous studies in aquaculture have emphasized microalgae as the ultimate "superfood" for aquatic life [80–82]. In this context, the protein, lipid, and carbohydrate content of plankton (phyto- and zooplankton) is highlighted concerning the most popular commercial aquafeed ingredients, as presented in **Figure 1** and listed in **Table 2** (Ashour [72], Elshobary and Ashour [73], and Mansour et al. [74]). Plankton is the premier source of minerals, nutrients, vitamins, lipids, PUFA, MUFA, phospholipids (PLs), carbohydrate, monosaccharides, polysaccharides, protein, EAAs, non-EAAs, and peptides. Phytoplankton contains pigments, such as chlorophylls, phycocyanin, xanthophylls, lutein, carotenoid, astaxanthin, phycobiliproteins, and phytol, as well as physiologically active chemicals, such as hydrocarbons, alkaloids, terpenoids, and phenol with antibacterial and antioxidant characteristics [72, 104, 105].

Planktonic cells have diverse uses in aquaculture, spanning three main phases: (1) hatchery seeds/incubations, (2) early rearing/nurseries, and (3) on-growing/mass production [72]. Phytoplankton and zooplankton are key feed for various aquatic life stages, including larvae, post-larvae, juveniles, adult bivalves, mollusks, crustaceans, and all marine hatcheries [106]. In the aquaculture and hatcheries industry, phytoplankton is utilized in a variety of ways, including live food, pastes, dried biomass, spray-dried, freeze-dried, flakes, defatted biomass (from the creation of biodiesel), and trash from microalgae derived from the bioindustries [107, 108].

Microalgae, often known as phytoplankton, are the food chain's foundation in natural aquatic ecosystems [109]. Consequently, they play a crucial role in supporting the commercial cultivation of various aquatic animals, serving as a primary food source for all growth stages of bivalves, mollusks, sea cucumbers, seahorses, early crustacean larvae, and the initial development of certain fish species [110, 111]. While numerous microalgae species have been tested as live food in marine hatcheries, only about 10 species have explained widespread application in aquaculture [72, 112]. As presented in **Figure 1** and **Table 2**, within aquaculture systems and natural aquatic environments, phytoplanktonic organisms serve as fundamental components of the aquatic food web, meeting the nutritional needs of wild zooplankton species, such as rotifers, copepods, amphipods, and daphnia. These zooplankton species are then used as live feeds for crustaceans, fish, and fish larvae in a sequential order [105].

Item	Biochemical composition (%, dry weight bases)			References
	Protein	Carbohydrate	Lipid	
Phytoplankton species				
Chaetoceros muelleri	59.00	10.00	31.00	[83]
Chaetoceros calcitrans	40.00	37.00	3.00	[83]
Skeletonema costatum	25	4–6	10	[84]
Thalassiosira pseudonana	34.00	9.00	19.00	[84]
Isochrysis galbana	27	34	11	[85]
Nannochloropsis	18–34	27–36	24–28	[86–88]
Pavlova sp.	24–29	6–9	9–14	[84, 89]
Tetraselmis suecica	38.73	44.29	12.38	[90]
Chlorella vulgaris	51–58	2–17	14–22	[89]
Haematococcus pluvialis	17–45	20–37	15–40	[91]
Dunaliella salina	11–34	14–32	6–14	[85, 92]
Zooplankton species				
Rotifer, *Brachionus plicatilis*	52.9–54.4	32.8–34.5	4.5–6.27	[93]
Rotifer, *B. plicatilis*	29.2–38.3	11.3–14.4	15.9–35.1	[94]
Rotifer, *B. plicatilis*	44.6–63.6	—	22.2–32.5	[95]
Artemia sp. (newly hatched nauplii)	49.1–62.6	—	12.5–24.8	[95]
Artemia salina (strain San Francisco Bay)	47.2–61.2	10.3–12.1	21.9–28.5	[96]
Artemia salina (strain Great Salt Lake)	47.2–64.9	8.3–19.4	18.5–22.5	[96]
Copepod, *Euterpina acutifrons*	28.5–41.9	6.3–9.5	1.8–5.6	[97]
Copepod, *Acartia spinicauda*	67.3–74.9	6.6–9.7	12.8–17.2	[98]
Copepod, *Oithona similis*	59.9–70.1	3.4–6.2	11.6–14.8	[98]
Commercially aquafeed ingredients				
Saccharomyces cerevisiae	50	1.8	4.6	[99]
Soybean	44	2.2	39	[100]
Fish meal	63	11	12.5	[101]
Wheat meal	12.2	2.9	69	[102]
Corn-gluten meal	62	5	18.5	[103]

Table 2.
The typical nutritional profiles (protein, lipid, and carbohydrate) of phytoplankton and zooplankton, compared to those of the most commercially aquafeed ingredients.

In marine hatcheries, phytoplankton is farmed to feed zooplankton, such as rotifers, copepods, and *Artemia*, which then provide as food for fish and shellfish [113–116].

4.2 Zooplankton

Numerous zooplankton species, such as rotifers (*Brachionus* sp.), copepods (calanoid, cyclopoid, and harpacticoid), cladocerans (*Moina* sp., *Daphnia* sp., and *Ceriodaphnia* sp.), larval forms of brine shrimp (*Artemia salina* and *Artemia* sp.), and larval forms of various aquatic organisms, have been mass-cultured for hatchery utilization and are thought to be a natural food source for fish and shellfish [117]. In marine hatcheries, rotifers, *Artemia*, and copepods are the most significant zooplankton live feed species [118, 119]. Regarding rotifers, *Brachionus plicatilis* and *Brachionus rotundiformis* are the two rotifer species informed worldwide [120]. The two species of *Artemia* that are most commonly and commercially utilized in marine hatcheries for the development of marine larvae are *Artemia franciscana* and *A. salina* [121]. Additionally, several copepod species, including *Oithona nana*, *Acartia* spp., *Oithona rigida*, *Bestiolina* sp., *Temora stylifera*, *Nannocalanus minor*, and *Paracalanus pas*, have also been commercialized [122–130].

Rotifers are a category of small zooplankton with notable traits, such as slow movement, compact size, environmental adaption, the ability to float in water, and a high rate of reproduction when compared to many other zooplankton species [131]. They can also be supplemented with fatty acids and antibiotics, and they are rather simple to cultivate in huge quantities. In hatcheries around the world, euryhaline rotifers, such as *Brachionus* spp., are especially important for the intense rearing of marine larval finfish [132]. In aquaculture, only a few rotifer species from the *Brachionus* genus—*Brachionus plicatilis* (L-strain), which ranges from 200 to 360 μm, and *B. rotundiformis* (S-strain), which ranges from 150 to 220 μm—are used [133]. Rotifers actively consume particles in the water column that are usually between 1 and 10 μm in size. Rotifers are supposed to benefit from feeding a wide range of microalgae, including *Nannochloropsis* and *Isochrysis* [133]. Furthermore, various widely available artificial feeds are suitable for rotifer production, including yeast and algal diets. For example, activated baker's yeast has been efficiently employed as an affordable growth-promoting diet, with an optimal feeding rate of about 0.5 g/million rotifers [134]. On the other hand, it has been demonstrated that mixing microalgae with yeast increases growth rates. It is crucial to stress that to obtain most of the benefits from a diet deficient in critical fatty acids, amino acids, and certain vitamins, supplements are required [135]. During larviculture, the rotifer is often used as the main live feed when larvae shift from searching out external sources of energy (exogenous feeding) to depending on internal energy stores (endogenous feeding) [133]. As a transitional feed after the rotifer stage, *Artemia*, or brine shrimp, is frequently used as a live feed before the larvae transition to a dry diet [136]. Brine shrimp are primordial crustaceans that belong to the Branchiopoda class. Their entire length typically ranges between 0.7 and 1.2 mm [137]. This marine organism's ability to live and thrive in a wide variety of salt concentrations is extraordinary [138]. *Artemia*, which has over 50 identified strains globally, is one of the most often utilized live feeds in aquaculture due to its accessibility and ease of use [139]. The ability of *Artemia* to create "cysts," or dormant embryos, is one of its primary traits. These cysts can be harvested, processed, preserved, or purchased commercially [140]. These cysts incubate in seawater for around 24 hours before hatching into free-swimming nauplii, which provide a nutrient-dense live food source for the larvae of many marine animals [139]. *Artemia* is the most practical and least labor-intensive live food choice for aquaculture because of this special quality [141]. Brine shrimp cysts are often acquired from commercial vendors and hatched in aquariums [141, 142]. Brine shrimp, like rotifers, must be supplemented, prior to feeding fish larvae. *Artemia* devour a wide range of foods since they are opportunistic feeders [139]. With over 21,000 species documented, copepods make up around 85% of sea-dwelling zooplankton and are an essential part of the marine zooplankton population [143, 144]. As an essential component of the nutrition flow from primary producers to marine fish larvae, these microscopic crustaceans are vital to marine ecosystems [145]. With larvae eating copepod nauplii and young fish feeding on adult copepods, they constitute an essential food supply for a large number of marine species. In contrast to rotifers and brine shrimp, copepods can naturally synthesize essential highly unsaturated fatty acids (HUFAs) without the requirement for enrichment. This permits them to maintain the DHA:EPA and EPA:ARA ratios, which are critical for marine fish larvae [146]. Calanoids are distinguished by their notably long first antennae (16–26 segments), while Harpacticoids possess shorter first antennae (less than 10 segments) [144]. Some copepod species from these groups are cultivated for hatchery operations, including Calanoids, such as *Acartia tonsa*, *Eurytemora affinis*, and *Pseudocalanus elongatus*, as well as Harpacticoids like *Tisbe holothuriae*, *Tigriopus japonicus,* and *Schizopera elatensis* [137, 147, 148].

In hatchery activities, as presented in **Table 2**, regarding the fact that the phytoplankton is a basic aquatic food chain [109], it provides all nutritional requirements to the zooplankton species that are sequentially utilized as live feeds, besides phytoplankton for fish and shellfish larvae [113–116]. Shrimp species rely on phytoplankton, in particular, on diatoms for essential nutrition during various developmental stages (nauplius, zoea, mysis, and post-larval stages). Bacillariophyceae (diatoms) are favored for shrimp due to their small cell size, rapid growth, and rich nutritional profile, including high PUFA, EPA, and DHA content [149]. Species like *Skeletonema* and *Chaetoceros* are crucial for shrimp larval development and rearing [150].

Additionally, species like *Tetraselmis*, *Dunaliella,* and *Isochrysis* are important for shrimp larvae due to their nutritional benefits [151]. In the life cycles of bivalves, phytoplankton, including *Tetraselmis* and *Chaetoceros*, and *Isochrysis,* significantly improve nutritional value and growth, development, and metamorphosis of bivalve [152–163]. In marine hatcheries, plankton (phyto and zoo) are introduced for marine fish and shellfish larvae in several forms, including live feeds, biomass (paste), free-lipid, and preserved concentrate [74].

Commercial rearing of bivalves mainly depends on the production of microalgae, which comprises about 30% of the operating costs of bivalve production in bivalve hatcheries. In marine bivalve hatcheries, a wide range of microalgal species, such as *Chaetoceros* sp., *Isochrysis galbana*, *Nitzschia* sp., *Phaeodactylum tricornutum*, *Pavlova* sp., *Skeletonema* sp., *Thalassiosira* sp., and *Tetraselmis* sp. [164], are used.

Sea cucumbers, prized for their commercial value, benefit from diets containing microalgae such as *Chaetoceros* and *Isochrysis*, which enhance growth and survival rates. Similarly, for sea urchins and seahorses, microalgae like *Isochrysis* and *Chaetoceros* are recommended for larvae, promoting survival, growth, and metamorphosis, especially the black-lip pearl oyster (*Pinctada margaritifera*) [165, 166].

Sea urchins and seahorses have a significant value as a seafood source, offering high nutritional, pharmacological, and therapeutic benefits. Despite their overfishing, efforts in hatching, larval rearing, and aquaculture present promising avenues for their sustainability and future development [167, 168]. Several studies highlight various phytoplankton species like *Isochrysis*, *Chaetoceros*, *Skeletonema,* and *Dunaliella* as perfect live feed species for the larvae development and growth of different sea urchin species such as *Tripneustes gratilla, Paracentrotus lividus*, and *Anthocidaris crassispina* [169–171].

Several studies indicate that applying *Nannochloropsis, Isochrysis*, *Chaetoceros,* and *Skeletonema* leads to significantly improved survival, ingestion, growth, and development of juvenile longsnout seahorses *Hippocampus reidi* [168, 172–175].

5. Conclusion and future perspective

Plankton (phytoplankton and zooplankton) are important primary food sources in marine hatcheries throughout the early stages of development, providing necessary nutrients and energy for growth and general health. Plankton's biochemical composition, which includes proteins, lipids, vitamins, minerals, and essential fatty acids, emphasizes its importance in supplying the nutritional needs of marine larvae in aquaculture conditions. Furthermore, the problems in creating and exploiting plankton in marine hatcheries indicate the need for future research and development in this area. Plankton generation is one of the most important components of marine hatcheries since it entails significant risk variables that render it non-impregnable. More than half of the costs associated with growing marine larvae are spent on producing live food in marine hatcheries. These issues have prompted extensive study in recent decades into a wide range of food sources, including microparticulate diets, micro-encapsulated and inert food, plankton paste, plankton free-lipid, and plankton concentrate stored in various ways. If aquaculture is to grow sustainably on a worldwide basis, it must still solve issues like live feed availability, nutritional value, and larval survival rates. In the future, it will be critical to conduct research on enhancing larval-rearing methods and plankton-based diet optimization. Technological developments can enhance the quality of live feed, which will help aquaculture productivity and larval professional development. Incorporating plankton-based diets into aquaculture promotes a sustainable sector that supports environmental preservation and food security by improving the growth and survival of marine larvae as well as helping to preserve wild populations.

Acknowledgements

The author would like to thank the UiT the Arctic University of Norway, Norway, for supporting the open access of this work.

Conflict of interest

The authors declare no conflict of interest.

Author details

Mohamed Ashour[1], Einar Ringø[2*], Ehab El-Haron[3] and Ashraf M.A. Goda[1]

1 National Institute of Oceanography and Fisheries (NIOF), Cairo, Egypt

2 Faculty of Biosciences, Fisheries, and Economics, Norwegian College of Fishery Science, UiT The Arctic University of Norway, Tromsø, Norway

3 Fish Nutrition Research Laboratory, Faculty of Agriculture, Animal Production Department, Cairo University, Cairo, Egypt

*Address all correspondence to: einar.ringo@uit.no

References

[1] Alam MA, Xu J-L, Wang Z. Microalgae Biotechnology for Food, Health and High Value Products. Singapore: Springer; 2020. DOI: 10.1007/978-981-15-0169-2

[2] Dhert P, Sorgeloos P. Live feeds in aquaculture. Infofish International. 1995;**2**:31-39

[3] Gisbert E, Luz RK, Fernández I, Pradhan PK, Salhi M, Mozanzadeh MT, et al. Development, nutrition, and rearing practices of relevant catfish species (Siluriformes) at early stages. Reviews in Aquaculture. 2022;**14**:73-105

[4] Steinberg CE. Aquatic Animal Nutrition: A Mechanistic Perspective from Individuals to Generations. Cham, Switzerland: Springer Nature; 2018. DOI: 10.1007/978-3-319-91767-2

[5] Drillet G, Jørgensen NO, Sørensen TF, Ramløv H, Hansen BW. Biochemical and technical observations supporting the use of copepods as live feed organisms in marine larviculture. Aquaculture Research. 2006;**37**:756-772

[6] Ma M, Hu Q. Microalgae as feed sources and feed additives for sustainable aquaculture: Prospects and challenges. Reviews in Aquaculture. 2024;**16**:818-835

[7] Conceição LE, Yúfera M, Makridis P, Morais S, Dinis MT. Live feeds for early stages of fish rearing. Aquaculture Research. 2010;**41**:613-640

[8] Estévez A, Papandroulakis N, Wille M, Sorgeloos P. Early life stages and weaning. In: Organic Aquaculture: Impacts and Future Developments. Cham, Switzerland: Springer Nature; 2019. pp. 79-102. DOI: 10.1007/978-3-030-05603-2_5

[9] Kandathil Radhakrishnan D, Akbar Ali I, Schmidt BV, John EM, Sivanpillai S, Thazhakot Vasunambesan S. Improvement of nutritional quality of live feed for aquaculture: An overview. Aquaculture Research. 2020;**51**:1-17

[10] Chen JY, Zeng C, Jerry DR, Cobcroft JM. Recent advances of marine ornamental fish larviculture: Broodstock reproduction, live prey and feeding regimes, and comparison between demersal and pelagic spawners. Reviews in Aquaculture. 2020;**12**:1518-1541

[11] Mohan K, Rajan DK, Muralisankar T, Ganesan AR, Sathishkumar P, Revathi N. Use of black soldier fly (*Hermetia illucens* L.) larvae meal in aquafeeds for a sustainable aquaculture industry: A review of past and future needs. Aquaculture. 2022;**553**:738095

[12] Rawski M, Mazurkiewicz J, Kierończyk B, Józefiak D. Black soldier fly full-fat larvae meal as an alternative to fish meal and fish oil in Siberian sturgeon nutrition: The effects on physical properties of the feed, animal growth performance, and feed acceptance and utilization. Animals. 2020;**10**:2119

[13] Radhakrishnan DK, Kumar S, Ali IA. Different animal feeds and their role in aquaculture. In: Aquaculture Science and Engineering. Singapore: Springer Nature; 2022. pp. 99-129. DOI: 10.1007/978-981-19-0817-0_4

[14] Costas-Selas C, Martínez-García S, Delgadillo-Nuno E, Justel-Díez M, Fuentes-Lema A, Fernández E, et al. Linking the impact of bacteria on phytoplankton growth with microbial community composition and co-occurrence patterns. Marine Environmental Research. 2024;**193**:106262

[15] Kandathil Radhakrishnan D, Velayudhannair K, Schmidt BV. Effects of bio-flocculated algae on the growth, digestive enzyme activity and microflora of freshwater fish Catla catla (Hamilton 1922). Aquaculture Research. 2020;**51**:4533-4540

[16] Bakhtiyar Y, Arafat MY, Andrabi S, Tak HI. Zooplankton: The significant ecosystem service provider in aquatic environment. In: Bioremediation and Biotechnology, Vol 3: Persistent and Recalcitrant Toxic Substances. Switzerland, AG: Springer Nature; 2020. pp. 227-244. DOI: 10.1007/978-3-030-46075-4

[17] Li Y, Xu Z, Liu H. Nutrient-imbalanced conditions shift the interplay between zooplankton and gut microbiota. BMC Genomics. 2021;**22**:1-18

[18] Ramlee A, Chembaruthy M, Gunaseelan H, Yatim S, Taufek H, Rasdi N. Enhancement of nutritional value on zooplankton by alteration of algal media composition: A review. Proceedings of IOP Conference Series: Earth and Environmental Science. 2021;**869**:012006

[19] Overstreet RM. Parasitic diseases of fishes and their relationship with toxicants and other environmental factors. In: Pathobiology of Marine and Estuarine Organisms. Boca Raton, Florida, United States: CRC Press, Taylor & Francis; 2021. pp. 111-156

[20] McEdward L. Ecology of Marine Invertebrate Larvae. Boca Raton, Florida, United States: CRC Press, Taylor & Francis; 2020

[21] Garrido S, Albo-Puigserver M, Moyano M. Larval trophic ecology of small pelagic fishes: A review of recent advances and pathways to fill remaining knowledge gaps. Marine Ecology Progress Series. 2024:**741**:127-143. DOI: 10.3354/meps14543

[22] Giri SS, Kim HJ, Jung WJ, Lee SB, Joo SJ, Gupta SK, et al. Probiotics in addressing heavy metal toxicities in fish farming: Current progress and perspective. Ecotoxicology and Environmental Safety. 2024;**282**:116755

[23] Kumar V, Mitra A, Ragaza JA. Digestive physiology of marine larvae: An overview of adaptations and mechanisms. In: Marine Larvae: Developments and Applications, by BySe-Kwon Kim. Taylor and Francis Group. 2024. pp. 251-284. DOI: 10.1201/9781003359388

[24] Kariyawasam K, Kariyawasam K, Rathnayake A, Wickramasinghe I. Nutrition of marine larvae. In: Marine Larvae. Boca Raton, Florida, United States: CRC Press, Taylor & Francis; 2024. pp. 126-157

[25] Agh N, Bossier P, Bosteels T, Boyko EG, Brown P, Camara MR, et al. Manual on Artemia production and use. In: FAO Fisheries and Aquaculture Technical Paper 2024. Rome: FAO; 2024. pp. I-190

[26] Hamre K, Yúfera M, Rønnestad I, Boglione C, Conceição L, Izquierdo M. Fish larval nutrition and feed formulation: Knowledge gaps and bottlenecks for advances in larval rearing. In: Success Factors for Fish Larval Production. Hoboken, New Jersey, United States: Wiley online Library; 2018. pp. 200-280

[27] Glencross BD. Exploring the nutritional demand for essential fatty acids by aquaculture species. Reviews in Aquaculture. 2009;**1**:71-124

[28] Parrish CC. Lipids in marine ecosystems. International Scholarly Research Notices. 2013;**2013**:604045

[29] Langdon C, Waldock M. The effect of algal and artificial diets on the growth and fatty acid composition of Crassostrea gigas spat. Journal of the Marine Biological Association of the United Kingdom. 1981;**61**:431-448

[30] Gurr GM, Wratten SD, Landis DA, You M. Habitat management to suppress pest populations: Progress and prospects. Annual Review of Entomology. 2017;**62**:91-109

[31] Tocher DR. Fatty acid requirements in ontogeny of marine and freshwater fish. Aquaculture Research. 2010;**41**:717-732

[32] Mejri SC, Tremblay R, Audet C, Wills PS, Riche M. Essential fatty acid requirements in tropical and cold-water marine fish larvae and juveniles. Frontiers in Marine Science. 2021;**8**:680003

[33] Tocher DR, Bendiksen EÅ, Campbell PJ, Bell JG. The role of phospholipids in nutrition and metabolism of teleost fish. Aquaculture. 2008;**280**:21-34

[34] Øie M, Sundet K, Ueland T. Neurocognition and functional outcome in early-onset schizophrenia and attention-deficit/hyperactivity disorder: A 13-year follow-up. Neuropsychology. 2011;**25**:25

[35] Drillet G, Lombard F. A first step towards improving copepod cultivation using modelling: The effects of density, crowding, cannibalism, tank design and strain selection on copepod egg production yields. Aquaculture Research. 2015;**46**:1638-1647

[36] Seychelles L, Audet C, Tremblay R, Fournier R, Pernet F. Essential fatty acid enrichment of cultured rotifers (*Brachionus plicatilis*, Müller) using frozen-concentrated microalgae. Aquaculture Nutrition. 2009;**15**:431-439

[37] Betancor MH. Anzaldúa and "the new mestiza": A Chicana dives into collective identity. Language Value. 2012;**4**:38-55

[38] Dantagnan P, Bórquez A, Hernández A, Izquierdo M. Effect of EPA/DHA ratios on the growth and survival of Galaxias maculatus (Jenyns, 1842) larvae reared under different salinity regimes. Aquaculture Research. 2010;**41**:e239-e244

[39] Das P, Mandal SC, Bhagabati S, Akhtar M, Singh S. Important live food organisms and their role in aquaculture. Frontiers in Aquaculture. 2012;**5**:69-86

[40] Salze G, McLean E, Craig SR. Dietary taurine enhances growth and digestive enzyme activities in larval cobia. Aquaculture. 2012;**362**:44-49

[41] Li X, Zheng S, Wu G. Nutrition and functions of amino acids in fish. In: Amino Acids in Nutrition and Health: Amino Acids in the Nutrition of Companion, Zoo and Farm Animals. Cham: Springer Nature; 2021. pp. 133-168 DOI: 10.1007/978-3-030-54462-1_8

[42] de Hernandez Dios MA, Tovar-Ramírez D, Maldonado García D, Galaviz-Espinoza MA, Spanopoulos Zarco M, Maldonado García MC. Functional additives as a boost to reproductive performance in marine fish: A review. Fishes. 2022;**7**:262

[43] Izquierdo M, Fernandez-Palacios H, Tacon A. Effect of broodstock nutrition on reproductive performance of fish. Aquaculture. 2001;**197**:25-42

[44] Fei S, Ou M, Liu H, Zhang X, Luo Q, Li K, et al. Transcriptomics and metabolomics reveal the regulation

of reproductive performance and egg quality of blotched snakehead *Channa maculata* broodstock induced by a high protein diet. Aquaculture Reports. 2024;**34**:101902

[45] Torsabo D, Ishak SD, Noordin NM, Waiho K, Chu IKC, Abduh MY, et al. Optimizing reproductive performance in pangasius catfish broodstock: A review of dietary and molecular strategies. Veterinary and Animal Science. 2024:100375. DOI: 10.1016/j.vas.2024.100375

[46] Gop AP, Ranjan R, Santhosh B, Surya S, Anil M. Broodstock Management of Marine Fish in RAS Tanks. Training Manual on Fisheries Management Practices and Techniques. Kochi: ICAR-Central Marine Fisheries Research Institute; 2024. p. 154

[47] Boidron-Métairon IF. Larval nutrition. In: Ecology of Marine Invertebrate Larvae. Boca Raton, Florida, United States: CRC Press, Taylor & Francis; 2020. pp. 223-248

[48] Li X, Han T, Zheng S, Wu G. Nutrition and functions of amino acids in aquatic crustaceans. Advances in Experimental Medicine and Biology. 2021;**1285**:169-198

[49] Samat NA, Yusoff FM, Rasdi NW, Karim M. Enhancement of live food nutritional status with essential nutrients for improving aquatic animal health: A review. Animals. 2020;**10**:2457

[50] Williams KC. Nutritional requirements and feeds development for post-larval spiny lobster: A review. Aquaculture. 2007;**263**:1 14

[51] Rathore S, Yusufzai S, Katira N, Jaiswal K. Fish larval nutrition: A review on new developments. The International Journal of Engineering and Science. 2016;**9**:40-47

[52] Morais S, Koven W, Rønnestad I, Dinis MT, Conceição LE. Dietary protein/lipid ratio affects growth and amino acid and fatty acid absorption and metabolism in Senegalese sole (*Solea senegalensis* Kaup 1858) larvae. Aquaculture. 2005;**246**:347-357

[53] Morais S, Koven W, Rønnestad I, Dinis MT, Conceição LE. Dietary protein: Lipid ratio and lipid nature affects fatty acid absorption and metabolism in a teleost larva. British Journal of Nutrition. 2005;**93**:813-820

[54] Yúfera M, Fernández-Díaz C, Pascual E. Food microparticles for larval fish prepared by internal gelation. Aquaculture. 2005;**248**:253-262

[55] Rubio V, Sánchez-Vázquez F, Madrid J. Macronutrient selection through postingestive signals in sea bass fed on gelatine capsules. Physiology & Behavior. 2003;**78**:795-803

[56] Lall SP, Kaushik SJ. Nutrition and metabolism of minerals in fish. Animals. 2021;**11**:2711

[57] Hamre K, Yufera M, Rønnestad I, Boglione C, Conceição LE, Izquierdo M. Fish larval nutrition and feed formulation: Knowledge gaps and bottlenecks for advances in larval rearing. Reviews in Aquaculture. 2013;**5**:S26-S58

[58] Hamre K, Holen E, Moren M. Pigmentation and eye migration in Atlantic halibut (*Hippoglossus hippoglossus* L.) larvae: New findings and hypotheses. Aquaculture Nutrition. 2007;**13**:65-80

[59] Van Tien Nguyen VTN, Satoh S, Haga Y, Fushimi H, Kotani T. Effect of zinc and manganese supplementation

in Artemia on growth and vertebral deformity in red sea bream (*Pagrus major*) larvae. Aquaculture. 2008;**285**:184-192

[60] Penglase S, Hamre K, Sweetman J, Nordgreen A. A new method to increase and maintain the concentration of selenium in rotifers (*Brachionus* spp.). Aquaculture. 2011;**315**:144-153

[61] Ribeiro A, Ribeiro L, Sæle Ø, Hamre K, Dinis M, Moren M. Iodine-enriched rotifers and Artemia prevent goitre in Senegalese sole (*Solea senegalensis*) larvae reared in a recirculation system. Aquaculture Nutrition. 2011;**17**:248-257

[62] Hamre K, Harboe T. Artemia enriched with high n-3 HUFA may give a large improvement in performance of Atlantic halibut (*Hippoglossus hippoglossus* L.) larvae. Aquaculture. 2008;**277**:239-243

[63] Council, NR. Nutrient Requirements of Fish and Shrimp. Ottawa, Ontario, Canada: National Research Council; 2011

[64] Busch KET, Falk-Petersen IB, Peruzzi S, Rist NA, Hamre K. Natural zooplankton as larval feed in intensive rearing systems for juvenile production of Atlantic cod (*Gadus morhua* L.). Aquaculture Research. 2010;**41**:1727-1740

[65] Hamre K, Mollan TA, Sæle Ø, Erstad B. Rotifers enriched with iodine and selenium increase survival in Atlantic cod (*Gadus morhua*) larvae. Aquaculture. 2008;**284**:190-195

[66] Penglase S, Nordgreen A, Van der Meeren T, Olsvik P, Sæle Ø, Sweetman J, et al. Increasing the level of selenium in rotifers (*Brachionus plicatilis* 'Cayman') enhances the mRNA expression and activity of glutathione peroxidase in cod (*Gadus morhua* L.) larvae. Aquaculture. 2010;**306**:259-269

[67] Jobling M, National Research Council (NRC). Nutrient Requirements of Fish and Shrimp. Washington, DC: The National Academies Press. Springer; 2011, 2012. p. 376, ISBN: 978-0-309-16338-5

[68] Kamler E. Resource allocation in yolk-feeding fish. Reviews in Fish Biology and Fisheries. 2008;**18**:143-200

[69] Rønnestad I, Hamre K, Lie Ø, Waagbø R. Ascorbic acid and α-tocopherol levels in larvae of Atlantic halibut before and after exogenous feeding. Journal of Fish Biology. 1999;**55**:720-731

[70] Rønnestad I, Hemre G-I, Finn R, Lie Ø. Alternate sources and dynamics of vitamin A and its incorporation into the eyes during the early endotrophic and exotrophic larval stages of Atlantic halibut (*Hippoglossus hippoglossus* L.). Comparative Biochemistry and Physiology Part A: Molecular & Integrative Physiology. 1998;**119**:787-793

[71] Rønnestad I, Lie Ø, Waagbø R. Vitamin B6 in Atlantic halibut, *Hippoglossus hippoglossus*—Endogenous utilization and retention in larvae fed natural zooplankton. Aquaculture. 1997;**157**:337-345

[72] Ashour M. Current and future perspectives of microalgae-aquaculture in Egypt, case study: SIMAF-prototype-project. Egyptian Journal of Animal Production. 2020;**57**:163-170

[73] Elshobary M, Ashour M. Algae for aquaculture: Recent technological applications. In: Value-Added Products from Algae: Phycochemical Production and Applications. Cham, Switzerland

AG: Springer Nature; 2023. pp. 525-561. DOI: 10.1007/978-3-031-42026-9

[74] Mansour AT, Ashour M, Alprol AE, Alsaqufi AS. Aquatic plants and aquatic animals in the context of sustainability: Cultivation techniques, integration, and blue revolution. Sustainability. 2022;**14**:3257

[75] Brown MR. Nutritional value and use of microalgae in aquaculture. In: Advances en Nutrición Acuicola. Australia: CSIRO Marine Research, CSIRO Publishing, Clayton; 2002

[76] Ashour M, Omran AM. Recent advances in marine microalgae production: Highlighting human health products from microalgae in view of the coronavirus pandemic (COVID-19). Fermentation. 2022;**8**:466

[77] Cembella AD, Antia NJ, Harrison PJ. The utilization of inorganic and organic phosphorous compounds as nutrients by eukaryotic microalgae: A multidisciplinary perspective: Part I. CRC Critical Reviews in Microbiology. 1982;**10**:317-391

[78] Oostlander P, van Houcke J, Wijffels RH, Barbosa MJ. Microalgae production cost in aquaculture hatcheries. Aquaculture. 2020;**525**:735310

[79] Ashour M, Kamel A. Enhance growth and biochemical composition of *Nannochloropsis oceanica*, cultured under nutrient limitation, using commercial agricultural fertilizers. Journal of Marine Science: Research and Development. 2017;7:1-5

[80] FAO. The State of World Fisheries and Aquaculture 2020. Sustainability in Action. Rome: FAO; 2020. DOI: 10.4060/ca9229en

[81] Barra L, Chandrasekaran R, Corato F, Brunet C. The challenge of ecophysiological biodiversity for biotechnological applications of marine microalgae. Marine Drugs. 2014;**12**:1641-1675

[82] Ferdouse F, Holdt SL, Smith R, Murúa P, Yang Z. The global status of seaweed production, trade and utilization. Globefish Research Programme. 2018;**124**:I

[83] Velasco LA, Carrera S, Barros J. Isolation, culture and evaluation of *Chaetoceros muelleri* from the Caribbean as food for the native scallops, *Argopecten nucleus* and *Nodipecten nodosus*. Latin American Journal of Aquatic Research. 2016;**44**:557-568

[84] Brown MR. The amino-acid and sugar composition of 16 species of microalgae used in mariculture. Journal of Experimental Marine Biology and Ecology. 1991;**145**:79-99

[85] da Silva Gorg CM, Aranda DAG. Morphological and chemical aspects of *Chlorella pyrenoidosa*, *Dunaliella tertiolecta*, *Isochrysis galbana* and *Tetraselmis gracilis microalgae*. Natural Science. 2013;**5**(7):33918, 9. DOI: 10.4236/ns.2013.57094

[86] Niccolai A, Zittelli GC, Rodolfi L, Biondi N, Tredici MR. Microalgae of interest as food source: Biochemical composition and digestibility. Algal Research. 2019;**42**:101617

[87] Tibbetts SM, Milley JE, Lall SP. Chemical composition and nutritional properties of freshwater and marine microalgal biomass cultured in photobioreactors. Journal of Applied Phycology. 2015;**27**:1109-1119

[88] Ashour M, Elshobary ME, El-Shenody R, Kamil A-W, Abomohra AE-F. Evaluation of a native oleaginous marine microalga *Nannochloropsis*

oceanica for dual use in biodiesel production and aquaculture feed. Biomass and Bioenergy. 2019;**120**:439-447

[89] Becker EW. Micro-algae as a source of protein. Biotechnology Advances. 2007;**25**:207-210

[90] Sharawy ZZ, Ashour M, Abbas E, Ashry O, Helal M, Nazmi H, et al. Effects of dietary marine microalgae, *Tetraselmis suecica*, on production, gene expression, protein markers and bacterial count of Pacific white shrimp *Litopenaeus vannamei*. Aquaculture Research. 2020;**51**:2216-2228

[91] Shah M, Mahfuzur R, Liang Y, Cheng JJ, Daroch M. Astaxanthin-producing green microalga *Haematococcus pluvialis*: From single cell to high value commercial products. Frontiers in Plant Science. 2016;7:531

[92] Becker EW. Microalgae: Biotechnology and Microbiology. Vol. 10. Cambridge University Press; 1994

[93] Osmanoğlu Mİ, Elp M, Durmaz Y. Effects on fatty acids, biochemical composition and growth of rotifer (*Brachionus plicatilis*) fed with different concentrations of *Nannochloropsis* sp. Journal of Applied Phycology. 2024;**36**:1-9

[94] Jeeja P, Imelda J, Paulraj R. Nutritional composition of rotifer (*Brachionus plicatilis* Muller) cultured using selected natural diets. Indian Journal of Fisheries. 2011;**58**:59-65

[95] Naz M. The changes in the biochemical compositions and enzymatic activities of rotifer (*Brachionus plicatilis*, Müller) and Artemia during the enrichment and starvation periods. Fish Physiology and Biochemistry. 2008;**34**:391-404

[96] Claus C, Benijts F, Vandeputte G, Gardner W. The biochemical composition of the larvae of two strains of *Artemia salina* (L.) reared on two different algal foods. Journal of Experimental Marine Biology and Ecology. 1979;**36**:171-183

[97] Abdel Moati M, Atta M, El-Din N, Khalil A. Biochemical composition of the copepod *Euterpina acutifrons* from the coastal waters of alexandria. Qatar University Science Journal. 1994;**14**:395-403

[98] Perumal P, Rajkumar M, Santhanam P. Biochemical composition of wild copepods, *Acartia spinicauda* and *Oithona similis*, from *Parangipettai coastal* waters in relation to environmental parameters. Journal of Environmental Biology. 2009;**30**:995-1005

[99] Blomqvist J, Pickova J, Tilami SK, Sampels S, Mikkelsen N, Brandenburg J, et al. Oleaginous yeast as a component in fish feed. Scientific Reports. 2018;**8**:1-8

[100] A-FM E-S. Evaluation of soybean meal, spirulina meal and chicken offal meal as protein sources for silver seabream (*Rhabdosargus sarba*) fingerlings. Aquaculture. 1994;**127**:169-176

[101] Hodar A, Vasava R, Mahavadiya D, Joshi N. Fish meal and fish oil replacement for aqua feed formulation by using alternative sources: A review. Journal of Experimental Zoology-India. 2020;**23**:13-21

[102] Sørensen M, Morken T, Kosanovic M, Øverland M. Pea and wheat starch possess different processing characteristics and affect physical quality and viscosity of extruded feed for Atlantic salmon. Aquaculture Nutrition. 2011;**17**:e326-e336

[103] Liu W-Y, Fang L, Feng X-W, Li G-M, Gu R-Z. *In vitro* antioxidant and angiotensin I-converting enzyme inhibitory properties of peptides derived from corn gluten meal. European Food Research and Technology. 2020;**246**:2017-2027

[104] Kaparapu J. Application of microalgae in aquaculture. Phykos. 2018;**48**:21-26

[105] Wikfors GH, Ohno M. Impact of algal research in aquaculture. Journal of Phycology. 2001;**37**:968-974

[106] Yang W, Zheng Z, Lu K, Zheng C, Du Y, Wang J, et al. Manipulating the phytoplankton community has the potential to create a stable bacterioplankton community in a shrimp rearing environment. Aquaculture. 2020;**520**:734789

[107] Delgado E, Valles-Rosales DJ, Flores NC, Reyes-Jáquez D. Evaluation of fish oil content and cottonseed meal with ultralow gossypol content on the functional properties of an extruded shrimp feed. Aquaculture Reports. 2021;**19**:100588

[108] Cardinaletti G, Messina M, Bruno M, Tulli F, Poli B, Giorgi G, et al. Effects of graded levels of a blend of *Tisochrysis lutea* and *Tetraselmis suecica* dried biomass on growth and muscle tissue composition of European sea bass (*Dicentrarchus labrax*) fed diets low in fish meal and oil. Aquaculture. 2018;**485**:173-182

[109] Alprol AE, Ashour M, Mansour AT, Alzahrani OM, Mahmoud SF, Gharib SM. Assessment of water quality and phytoplankton structure of eight Alexandria beaches, southeastern Mediterranean Sea, Egypt. Journal of Marine Science and Engineering. 2021;**9**:1328

[110] Spolaore P, Joannis-Cassan C, Duran E, Isambert A. Commercial applications of microalgae. Journal of Bioscience and Bioengineering. 2006;**101**:87-96

[111] Duy NDQ, Francis DS, Southgate PC. The nutritional value of live and concentrated micro-algae for early juveniles of sandfish, *Holothuria scabra*. Aquaculture. 2017;**473**:97-104

[112] Brown M, Robert R. Preparation and assessment of microalgal concentrates as feeds for larval and juvenile Pacific oyster (*Crassostrea gigas*). Aquaculture. 2002;**207**:289-309

[113] Abo-Taleb H, Ashour M, El-Shafei A, Alataway A, Maaty MM. Biodiversity of *Calanoida copepoda* in different habitats of the North-Western Red Sea (Hurghada shelf). Water. 2020;**12**:656

[114] Alprol AE, Heneash AMM, Soliman AM, Ashour M, Alsanie WF, Gaber A, et al. Assessment of water quality, eutrophication, and zooplankton community in lake Burullus, Egypt. Diversity. 2021;**13**:268

[115] Ashour M, Abo-Taleb HA, Hassan A-KM, Abdelzaher OF, Mabrouk MM, Elokaby MA, et al. Valorization use of amphipod meal, *Gammarus pulex*, as a fishmeal substitute on growth performance, feed utilization, histological and histometric indices of the gut, and economic revenue of grey mullet. Journal of Marine Science and Engineering. 2021;**9**:1336

[116] Abo-Taleb HA, Ashour M, Elokaby MA, Mabrouk MM, El-feky MMM, Abdelzaher OF, et al. Effect of a new feed *Daphnia magna* (Straus, 1820), as a fish meal substitute on growth, feed utilization, histological status, and economic revenue of grey mullet, *Mugil*

cephalus (Linnaeus 1758). Sustainability. 2021;**13**:7093

[117] Brown M, Blackburn S. Live microalgae as feeds in aquaculture hatcheries. In: Advances in Aquaculture Hatchery Technology. Elsevier; 2013. pp. 117-158e

[118] Abdel-Rahim MM, Mansour AT, Mona MH, El-Gamal MM, El Atafy MM. To what extent can maternal inherited immunity acquired from a crustacean-enhanced diet improve the performance and vitality of the offspring and enhance profitability of European Sea bass (*Dicentrarchus labrax*)? Journal of the World Aquaculture Society. 2019;**50**:550-574

[119] El-Gamal MM, Othman SI, Abdel-Rahim MM, Mansour AT, Alsaqufi AS, El Atafy MM, et al. Palaemon and artemia supplemented diet enhances sea bass, *Dicentrarchus labrax*, broodstock reproductive performance and egg quality. Aquaculture Reports. 2020;**16**:100290

[120] Navarro N, Sarasquete C. Use of freeze-dried microalgae for rearing gilthead seabream, *Sparus aurata*, larvae: I. Growth, histology and water quality. Aquaculture. 1998;**167**:179-193

[121] Zaki M, Saad H. Comparative study on growth and survival of larval and juvenile *Dicentrarchus labrax* rearing on rotifer and Artemia enriched with four different microalgae species. African Journal of Biotechnology. 2010;**9**:3676-3688

[122] Camus T, Zeng C. Roles of microalgae on total egg production over female lifespan and egg incubation time, naupliar and copepodite survival, sex ratio and female life expectancy of the copepod *Bestiolina similis*. Aquaculture Research. 2010;**41**:1717-1726

[123] Ianora A, Romano G, Carotenuto Y, Esposito F, Roncalli V, Buttino I, et al. Impact of the diatom oxylipin 15S-HEPE on the reproductive success of the copepod *Temora stylifera*. Hydrobiologia. 2011;**666**:265-275

[124] Santhanam P, Jeyaraj N, Jothiraj K, Ananth S, Kumar SD, Pachiappan P. Assessing the efficacy of marine copepods as an alternative first feed for larval production of tiger shrimp *Penaeus monodon*. In: Basic and Applied Zooplankton Biology. Singapore: Springer Nature; 2019. pp. 293-303. DOI: 10.1007/978-981-10-7953-5_12

[125] Ananth S, Santhanam P. Intensive culture, biochemical composition analysis, and use of zooplankton *Tisbe* sp. (*Copepoda*: *Harpacticoida*) as an alternative live feed for shrimp larviculture. In: Basic and Applied Zooplankton Biology. Singapore: Springer; 2019. pp. 329-362. DOI: 10.1007/978-981-10-7953-5_15

[126] Jothiraj K, Santhanam P. Optimisation of the culture conditions of *Nannocalanus minor* (*Copepoda*: *Calanoida*). In: Basic and Applied Zooplankton Biology. Singapore: Springer; 2019. pp. 225-246. DOI: 10.1007/978-981-10-7953-5_8

[127] Schipp GR, Bosmans JM, Marshall AJ. A method for hatchery culture of tropical calanoid copepods, *Acartia* spp. Aquaculture. 1999;**174**:81-88

[128] Altaff K, Vijayaraj R. Micro-algal diet for copepod culture with reference to their nutritive value–A review. International Journal of Current Research and Review. 2021;**13**:86

[129] Magouz FI, Essa MA, Matter M, Tageldein Mansour A, Alkafafy M, Ashour M. Population dynamics, fecundity and fatty acid composition of

Oithona nana (*Cyclopoida*, *Copepoda*), fed on different diets. Animals. 2021;**11**:1188

[130] Magouz FI, Essa MA, Matter M, Mansour AT, Gaber A, Ashour M. Effect of different salinity levels on population dynamics and growth of the cyclopoid copepod *Oithona nana*. Diversity. 2021;**13**:190

[131] Wallace RL, Snell TW. Rotifera. In: Ecology and Classification of North American Freshwater Invertebrates. 3rd ed. USA: Elsevier; 2010. pp. 173-235. DOI: 10.1016/B978-0-12-374855-3.00008-X

[132] Jana B. Aquatic Sciences in the Tropics: Plankton, Animal Community and Productivity. Boca Raton, Florida, United States: CRC Press, Taylor & Francis; 2024

[133] Marini F. The Breeder's Net: The Rotifer and Rotifer Home Culture. Advanced Aquarist; 2002

[134] Shields R. Larviculture of marine finfish in Europe. Aquaculture. 2001;**200**:55-88

[135] Megarajan S, Chinnibabu B, Sadhu N. Rotifer Culture Techniques for Marine Finfish Larval Rearing. Kochi, India: CMFRI Repository, Library and AKMU of Central Marine Fisheries Research Institute; 2023

[136] Sorgeloos P, Coutteau P, Dhert P, Merchie GL. Use of brine shrimp, *Artemia* spp., in larval crustacean nutrition: A review. Reviews in Fisheries Science. 1998;**6**:55-68

[137] Lavens P, Sorgeloos P. The history, present status and prospects of the availability of Artemia cysts for aquaculture. Aquaculture. 2000;**181**:397-403

[138] Sorgeloos P, Dhert P, Candreva P. Use of the brine shrimp, *Artemia* spp., in marine fish larviculture. Aquaculture. 2001;**200**:147-159

[139] Gajardo GM, Beardmore JA. The brine shrimp Artemia: Adapted to critical life conditions. Frontiers in Physiology. 2012;**3**:185

[140] Ogburn NJ, Duan L, Subashchandrabose SR, Sorgeloos P, O'Connor W, Megharaj M, et al. Agricultural wastes for brine shrimp Artemia production: A review. Reviews in Aquaculture. 2023;**15**:1159-1178

[141] Megarajan S, Avadhanula RK, Suresh R. Artemia Culture Techniques for Marine Finfish Larval Rearing. Kochi, India: CMFRI Repository, Library and AKMU of Central Marine Fisheries Research Institute; 2023

[142] Treanor HB, Guy CS, Ilgen JE, Sealey WM, Dove AT, Webb MA. Survival and growth of larval pallid sturgeon are improved by a live diet. North American Journal of Aquaculture. 2024;**86**:332-339

[143] Böttger-Schnack R. Community structure and vertical distribution of cyclopoid copepods in the Red Sea: I. Central Red Sea, autumn 1980. Marine Biology. 1990;**106**:473-485

[144] Støttrup J, Norsker N. Production and use of copepods in marine fish larviculture. Aquaculture. 1997;**155**:231-247

[145] Støttrup JG. Production and nutritional value of copepods. In: Live Feeds in Marine Aquaculture. Wiley Blackwell; 2003. pp. 145-205

[146] Kattner G, Hagen W. Lipids in marine copepods: Latitudinal

characteristics and perspective to global warming. In: Lipids in Aquatic Ecosystems. New York, NY: Springer; 2009. pp. 257-280. DOI: 10.1007/978-0-387-89366-2_11

[147] Punnarak P, Jarayabhand P, Piumsomboon A. Cultivation of harpacticoid copepods (families *Harpacticidae* and *Laophontidae*) under selected environmental conditions. Agriculture and Natural Resources. 2017;**51**:278-285

[148] Drillet G, Frouël S, Sichlau MH, Jepsen PM, Højgaard JK, Joarder AK, et al. Status and recommendations on marine copepod cultivation for use as live feed. Aquaculture. 2011;**315**:155-166

[149] Rodríguez-Núñez K, Toledo-Aguero P. Fatty acids profile and nutritional composition of two tropical diatoms from the costa Rican Pacific coast. Grasas y Aceites. 2017;**68**:e209-e209

[150] Sivakumar N, Sundararaman M, Selvakumar G. Efficacy of micro algae and cyanobacteria as a live feed for juveniles of shrimp *Penaeus monodon*. African Journal of Biotechnology. 2011;**10**:11594-11599

[151] Rohani-Ghadikolaei K, Abdolalian E, Hojatollah F, Masoud G, Ng WK. The nutritional effect of *Isochrysis galbana* and *Chaetoceros muelleri* cultured with different seaweed extracts on the larval development, growth and survival of the marine shrimp, *Penaeus indicus*. Aquaculture Research. 2015;**46**:1444-1454

[152] Ponis E, Robert R, Parisi G, Tredici M. Assessment of the performance of Pacific oyster (*Crassostrea gigas*) larvae fed with fresh and preserved *Pavlova lutheri* concentrates. Aquaculture International. 2003;**11**:69-79

[153] Knuckey RM, Brown MR, Robert R, Frampton DM. Production of microalgal concentrates by flocculation and their assessment as aquaculture feeds. Aquacultural Engineering. 2006;**35**:300-313

[154] McCausland MA, Brown MR, Barrett SM, Diemar JA, Heasman MP. Evaluation of live microalgae and microalgal pastes as supplementary food for juvenile Pacific oysters (*Crassostrea gigas*). Aquaculture. 1999;**174**:323-342

[155] Heasman M, Diemar J, O'connor W, Sushames T, Foulkes L. Development of extended shelf-life microalgae concentrate diets harvested by centrifugation for *Bivalve molluscs*–a summary. Aquaculture Research. 2000;**31**:637-659

[156] Laing I, Verdugo CG. Nutritional value of spray-dried *Tetraselmis suecica* for juvenile bivalves. Aquaculture. 1991;**92**:207-218

[157] Laing I, Child AR, Janke A. Nutritional value of dried algae diets for larvae of Manila clam (*Tapes philippinarum*). Journal of the Marine Biological Association of the United Kingdom. 1990;**70**:1-12

[158] Laing I, Millican PF. Indoor nursery cultivation of juvenile bivalve molluscs using diets of dried algae. Aquaculture. 1992;**102**:231-243

[159] Boeing P, Escondido C. Use of spray-dried *Schizochytrium* sp. as a partial algal replacement for juvenile bivalves. Journal of Shellfish Research. 1997;**16**:284

[160] Langdon C, Önal E. Replacement of living microalgae with spray-dried diets for the marine mussel *Mytilus galloprovincialis*. Aquaculture. 1999;**180**:283-294

[161] Albentosa M, Pérez-Camacho A, Labarta U, Fernández-Reiriz MJ. Evaluation of freeze-dried microalgal diets for the seed culture of *Ruditapes decussatus* using physiological and biochemical parameters. Aquaculture. 1997;**154**:305-321

[162] Arney B, Liu W, Forster IP, McKinley RS, Pearce CM. Feasibility of dietary substitution of live microalgae with spray-dried *Schizochytrium* sp. or *spirulina* in the hatchery culture of juveniles of the Pacific geoduck clam (*Panopea generosa*). Aquaculture. 2015;**444**:117-133

[163] Militz TA, Braley RD, Southgate PC. Factors influencing the capacity for pediveliger larvae of the giant clam, *Tridacna noae*, to ingest and digest cells of microalgae concentrates. Aquaculture. 2021;**533**:736121

[164] Hemaiswarya S, Raja R, Ravi Kumar R, Ganesan V, Anbazhagan C. Microalgae: A sustainable feed source for aquaculture. World Journal of Microbiology and Biotechnology. 2011;**27**:1737-1746

[165] Mélo RCS, Santos LPDS, Brito APM, Gouveia ADA, Marçal C, Cavalli RO. Use of the microalga *Nannochloropsis occulata* in the rearing of newborn longsnout seahorse *Hippocampus reidi* (Syngnathidae) juveniles. Aquaculture Research. 2016;**47**:3934-3941

[166] Martínez-Fernández E, Southgate PC. Use of tropical microalgae as food for larvae of the black-lip pearl oyster *Pinctada margaritifera*. Aquaculture. 2007;**263**:220-226

[167] Liu H, Kelly MS, Cook EJ, Black K, Orr H, Zhu JX, et al. The effect of diet type on growth and fatty-acid composition of sea urchin larvae, I. *Paracentrotus lividus* (Lamarck, 1816) (Echinodermata). Aquaculture. 2007;**264**:247-262

[168] Willadino L, Souza-Santos LP, Mélo RC, Brito AP, Barros NC, Araújo-Castro CM, et al. Ingestion rate, survival and growth of newly released seahorse *Hippocampus reidi* fed exclusively on cultured live food items. Aquaculture. 2012;**360**:10-16

[169] Scholtz R, Bolton J, Macey B. Effects of different microalgal feeds and their influence on larval development in the white-spined sea urchin *Tripneustes gratilla*. African Journal of Marine Science. 2013;**35**:25-34

[170] Carboni S, Vignier J, Chiantore M, Tocher DR, Migaud H. Effects of dietary microalgae on growth, survival and fatty acid composition of sea urchin *Paracentrotus lividus* throughout larval development. Aquaculture. 2012;**324**:250-258

[171] Shigai X, Chuanxin Q, Zhenhua M, Gang Y, Jinhui S, Wanni P, et al. Effects of dietary microalgae on growth and survival of larval development of sea urchin (*Anthocidaris crassispina*). 南方水产科学. 2020;**16**:115-120

[172] Kasimala M, Mogos G, Negasi K, Bereket G, Abdu M, Melake H. Biochemical composition of selected seaweeds from intertidal shallow waters of southern Red Sea. Eritrea. 2020;**47**(12):3934-3941. DOI: 10.1111/are.12843

[173] Koldewey H. Syngnathid Husbandry in Public Aquariums 2005 Manual. London, UK: Project Seahorse; 2005

[174] Olivotto I, Avella MA, Sampaolesi G, Piccinetti C, Ruiz PN, Carnevali O. Breeding and rearing the longsnout seahorse *Hippocampus*

reidi: Rearing and feeding studies. Aquaculture. 2008;**283**:92-96

[175] Pham NK, Lin J. The effects of different feed enrichments on survivorship and growth of early juvenile longsnout seahorse, *Hippocampus reidi*. Journal of the World Aquaculture Society. 2013;**44**:435-446